Multifunctional Organic–Inorganic Halide Perovskite

Multifunctional Organic–Inorganic Halide Perovskite

Applications in Solar Cells, Light-Emitting Diodes, and Resistive Memory

edited by

Nam-Gyu Park

Hiroshi Segawa

JENNY STANFORD PUBLISHING

Published by

Jenny Stanford Publishing Pte. Ltd.
Level 34, Centennial Tower
3 Temasek Avenue
Singapore 039190

Email: editorial@jennystanford.com
Web: www.jennystanford.com

British Library Cataloguing-in-Publication Data
A catalogue record for this book is available from the British Library.

Multifunctional Organic–Inorganic Halide Perovskite: Applications in Solar Cells, Light-Emitting Diodes, and Resistive Memory

ISBN 978-981-4800-52-5 (Hardcover)
ISBN 978-1-003-27593-0 (eBook)

Contents

Preface

Electricity generated by photovoltaic solar cells is clean energy without carbon dioxide emission. Photovoltaic effect was first discovered in 1839 using electrochemical cell with semiconducting AgBr. Since then, various semiconductors have been developed for photovoltaics. Power conversion efficiency (PCE) of a solar cell, defined as the ratio of the output power generated by a solar cell and the input power from sunlight, increased from 4.5%, using a silicone semiconductor in 1954, to 29.1%, using a compound semiconductor of GaAs in 2021 under standard one-sun illumination, according to Best Research-Cell Efficiency Chart provided by the National Renewable Energy Laboratory (NREL). This indicates that materials for solar cells are of importance. Recently, organic–inorganic halide perovskite has received great attention because of high PCE over 25% when it is used as a light absorber in solar cell structures. Excellent optoelectronic properties of this perovskite, such as high absorption coefficient and long carrier lifetime, are attributed to such an unprecedentedly superb photovoltaic performance. Besides photovoltaic application, it has been found that organic–inorganic halide perovskites also demonstrate multifunctional behaviors in many other optoelectronic applications.

This book discusses not only photovoltaics but also other optoelectronic applications, such as ferroelectricity, resistive switching memory, and light-emitting diode, to provide a better understanding of the physico-chemical and optoelectronic properties of organic–inorganic halide perovskite. In Chapter 1, electronic structure, phase transition, and surface defects are described to understand the intrinsic properties of perovskite materials. Further in-depth discussion on electronic band structure and charge transport and recombination is presented in Chapter 2. To understand photovoltaic and/or optoelectronic properties of organic–inorganic halide perovskite, the excited state is important rather than the ground state in the dark. In Chapter 3, the optical excited-state properties of halide perovskite are described. In Chapter 4, the ferroelectric mechanism and experimental observations of the organic–inorganic halide perovskite have been described to understand the origin and mechanism of the ferroelectric behavior observed in it. Chapter 5 describes tandem devices comprising a perovskite solar cell as a top cell and a bottom cell with narrower bandgap semiconductor, typically perovskite/perovskite homo-tandem structure and hetero-tandem structures of perovskite/silicon and perovskite/CIGS. Tandem structure utilizing a perovskite solar cell has a potential for higher efficiency over 35%. In Chapter 6, perovskite-based resistive memory is described. Since halide (iodide) migration is well known in halide perovskite, this migration is beneficial to resistive switching memory. Regarding application of perovskite in solar cell and module, upscaling is an important topic. In Chapter 7, carbon-based large area technology as a potential upscaling methodology is described to provide a way for manufacturing stable solar modules. Perovskite light-emitting diodes receive higher attention because of their high external quantum efficiency along with color sharpness. In Chapter 8, the working principle of perovskite light-emitting diode is described. In addition, morphology control of perovskite film is described in detail because the morphology has a significant influence on the light-emitting property. Since this book covers the multifunctional properties of organic–inorganic halide perovskite, we hope it will contribute to a better understanding of the fundamentals of this material and in developing various optoelectronic applications using halide perovskite.

Nam-Gyu Park
Suwon, Republic of Korea

Hiroshi Segawa
Tokyo, Japan

Chapter 1

Theoretical Investigations on Organometal Halide Perovskite

Ryota Jono,[a] Jotaro Nakazaki,[b] and Hiroshi Segawa[a,b]
[a]*Research Center for Advanced Science and Technology, University of Tokyo, 4-6-1 Komaba, Meguro-ku, Tokyo 153-8904, Japan*
[b]*Department of General Systems Studies, Graduate School of Arts and Sciences, University of Tokyo, 3-8-1 Komaba, Meguro-ku, Tokyo 153-8902, Japan*
csegawa@mail.ecc.u–tokyo.ac.jp

1.1 Introduction

Organometal halide perovskite is expressed as ABX_3 in chemical formula, where the monovalent cation A and the divalent cation B interact with three equivalents of halogen anion X to keep the neutrality of the total electric charges. Methylammonium lead triiodide ($MAPbI_3$) perovskite is one of the fundamental ABX_3 materials for solar cells [1] because of panchromatic coloration [2], ambipolar carrier transport [3], and good carrier mobility [4]. Therefore, $MAPbI_3$ derivatives by ionic substitution have been widely investigated for controlling properties, especially bandgap, which is the energy difference between the conduction band minimum

Multifunctional Organic–Inorganic Halide Perovskite: Applications in Solar Cells, Light-Emitting Diodes, and Resistive Memory
Edited by Nam-Gyu Park and Hiroshi Segawa

ISBN 978-981-4800-52-5 (Hardcover), 978-1-003-27593-0 (eBook)
www.jennystanford.com

(CBM) and the valence band maximum (VBM). Additionally, the surface and interface related to perovskite materials are important for high photoelectron conversion. Theoretical investigations of these issues have been performed to elucidate the optical and interfacial properties so far. In this chapter, we introduce some of the theoretical and computational analyses reported in this field.

1.2 Electronic Structure

The VBM of $MAPbI_3$ is the antibonding molecular orbital (MO) composed of Pb(6s) and I(5p), while the CBM of $MAPbI_3$ is mainly a nonbonding MO composed of Pb(6p) with a small antibonding MO composed of Pb(6p) and I(5p) [3, 5, 6]. The picture derived from nonrelativistic quantum chemistry is not adequate for the description of anisotropic electronic orbitals of a heavy atom. Therefore, spin-orbit coupling (SOC) should be considered to calculate the electronic structure of lead halide perovskite. The actual degree of SOC in organometal halide perovskite is well studied in the early stages of the study of the perovskite solar cells [7–25]. The bandgap is strongly affected by SOC acting on the CBM, that is Pb(6p). The CBM, which is triply degenerated in nonrelativistic representation, is split by SOC into twofold and fourfold degenerated states [6]. The SOC also generates a shift of CBM in *k*-space. This Rashba effect was discussed to explain the slow electron-hole recombination [26–29].

The degree of SOC also depends on the structure of the PbI_3^- framework. One of the most important roles of the A-site cation is the deformation of the PbI_3^- framework [30, 31]. As expressed in the Goldschmidt tolerance factor, the size of the A-site cation to form the perovskite structure is limited [32]. A small and/or anisotropic structure of the A-site cation will deform the crystal phase from cubic to tetragonal or orthorhombic. As a result, the energy level of the CBM is changed by the SOC depending on the octahedral tilting angle [30, 31].

The effective masses of electron or hole as photocarriers are one of the useful properties derived from the electronic band structure and expressed as follows:

$$m = \left(\frac{1}{\hbar^2} \frac{\partial^2 E(k)}{\partial k_i \partial k_j} \right)^{-1}, \tag{1.1}$$

where $\hbar$ is Planck constant and $E(k)$ is the band energy of an electron as a function of the wavevector k. One of the calculated electronic band structures for $MAPbI_3$ is shown in Fig. 1.1. The reported effective masses of cubic $MAPbI_3$ are 0.23 m_0 for electrons and 0.29 m_0 for holes and can explain experimentally the observed long-range ambipolar transport property of perovskite [3].

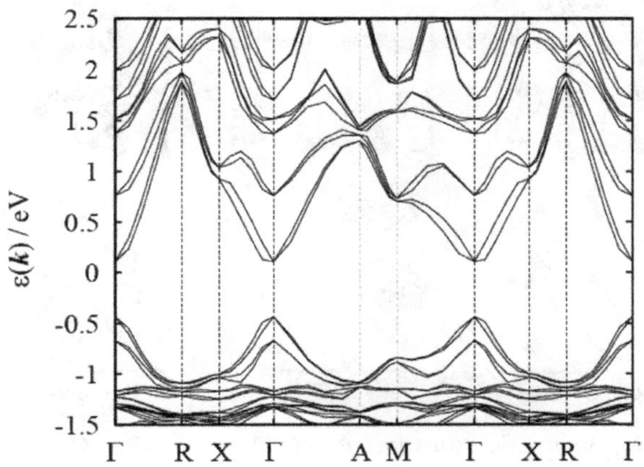

Figure 1.1 Electronic band structure of tetragonal $MAPbI_3$. PBEsol density functional with SOC is used for the calculation.

1.3 Phase Transitions

$MAPbI_3$ has three structural phases: the cubic $Pm\overline{3}m$ phase ($T > 327$ K), the tetragonal $I4/mcm$ phase (162 K $< T < 327$ K), and the orthorhombic $Pnma$ phase ($T < 162$ K) [33]. Their geometric structures are shown in Fig. 1.2.

The volume–potential energy plot calculated for each phase showed that the orthorhombic, tetragonal, and cubic phases are stable in smaller, medium, and bigger volumes, respectively, as shown in Fig. 1.3. The orthorhombic $Pnma$ phase is more stable than the tetragonal phase if the ratio of the lattice parameter c/a is larger than 1.45 [34]. When the in-plane lattice constant a increases, the tetragonal phase is stabilized because the orientation of the MA cation is changed from the [110] direction to the [100] direction. The cubic $Pm\overline{3}m$ phase is also stable at a larger volume because the

MA cation becomes oriented in the [111] direction [35]. The effect of the molecular motion of the MA cation in the cubic lead triiodide framework has also been discussed [36–39].

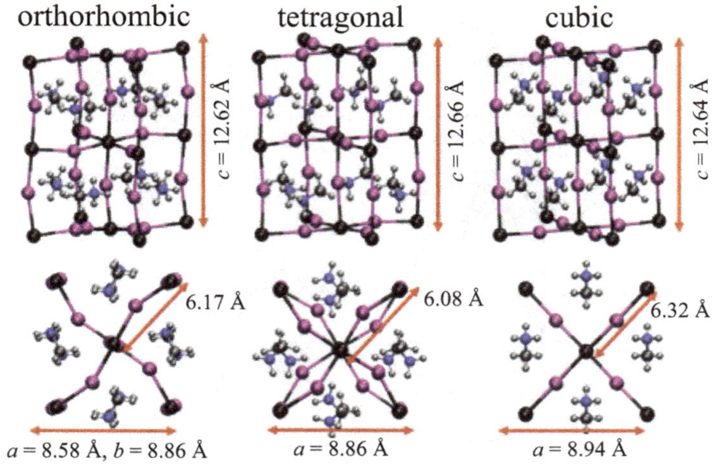

Figure 1.2 Geometric structure of $MAPbI_3$ perovskite. All structures are shown as $\sqrt{2}\times\sqrt{2}\times2$ supercells for comparison although the unit cell of the orthorhombic and cubic phases are $\sqrt{2}\times\sqrt{2}\times1$ and $1\times1\times1$, respectively.

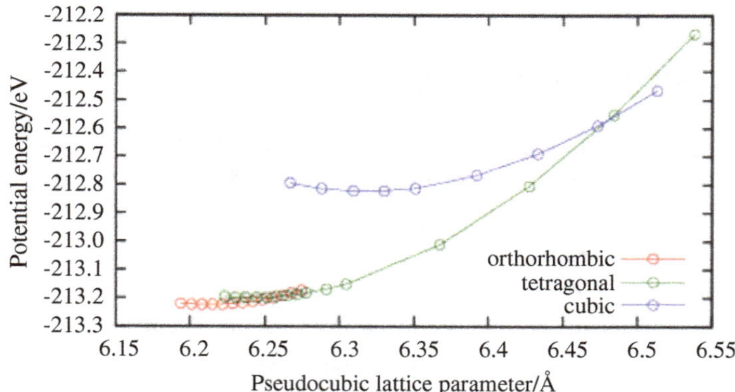

Figure 1.3 Potential energy as a function of the pseudocubic lattice parameter.

The phase transition of formamidinium lead triiodide ($FAPbI_3$) perovskite is also discussed. The transition pathway from the cubic $FAPbI_3$ perovskite structure (black) to the hexagonal $FAPbI_3$

structure (yellow) is simulated as shown in Fig. 1.4. It indicates that the energy barrier of this transition is about 600 meV/chemical formula unit and the formation enthalpy of the hexagonal phase is 70 meV lower than that of the cubic phase [40].

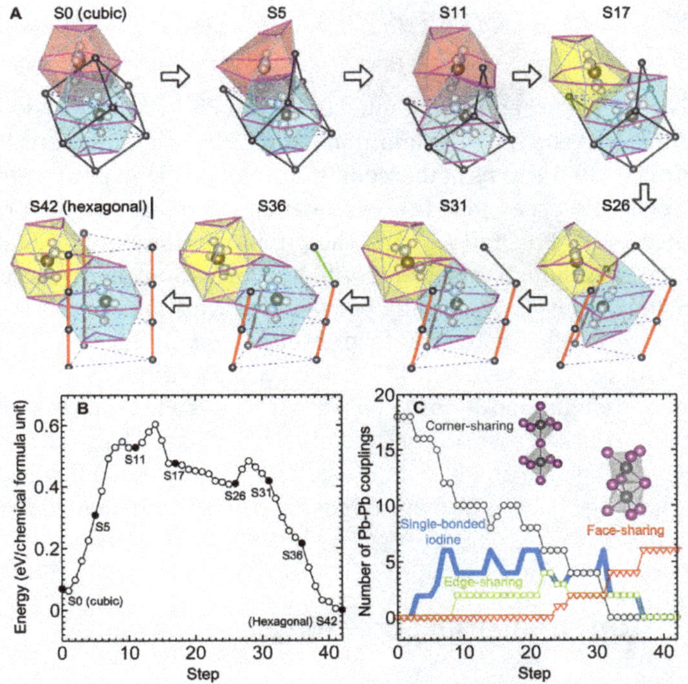

Figure 1.4 Transition pathway and energy barrier of the proposed reaction path from cubic $FAPbI_3$ perovskite to hexagonal $FAPbI_3$ material. Reprinted from Fig. 4 from Ref. [40] under a Creative Commons Attribution Non-commercial License 4.0 (CC BY-NC).

1.4 Surface, Interface, and Defects

The point defect in a semiconductor is a potential charge recombination center and should be controlled to obtain solar cells of a higher open-circuit voltage. The point defects are usually discussed by using the plot of the defect formation energy as a function of Fermi energy. The equilibrium growth condition of $MAPbI_3$ requires the following equation:

$$\mu_{\mathrm{MA}} + \mu_{\mathrm{Pb}} + 3\mu_{\mathrm{I}} = \Delta H(\mathrm{MAPbI_3}), \quad (1.2)$$

where μ_{MA}, μ_{Pb}, and μ_I are the chemical potentials of MA, Pb, and I. To exclude the formation condition of undesired products PbI_2 and MAI, the additional inequalities should be considered.

$$\mu_{\mathrm{Pb}} + 2\mu_{\mathrm{I}} < \Delta H(\mathrm{PbI_2}) \quad (1.3)$$

$$\mu_{\mathrm{MA}} + \mu_{\mathrm{I}} < \Delta H(\mathrm{MAI}) \quad (1.4)$$

The allowed chemical potential of Pb and I is very narrow, which indicates that the growth conditions of $MAPbI_3$ should be carefully controlled. On the basis of this condition, many types of point defects have been discussed [41–44]. The dominant defects are the *p*-type vacancy at the Pb site (V_{Pb}) and the *n*-type intersititial MA defects (MA_i) [41]. Although the defects with low formation energies create only shallow levels, the iodide or lead vacancies introduced in the surface of $MAPbI_3$ generate deep level defects by forming Pb–Pb or I–I dimers, respectively [45]. However, the deep level defects are easily eliminated by incorporation of a graphene layer on top of the $MAPbI_3$ perovskite because the interfacial charge transfer between graphene and dimers breaks the bonds formed by defects. This "healing effect" is also observed when using some small organic compounds [45].

1.5 Combinatorial Screening

The replacement of lead by nontoxic elements is one of big issues. Computational combinatorial screening using massively parallel many-core supercomputers has been performed to this end. ABX_3 and $A_2BB'X_6$ chemical formula are considered in these studies [46–48]. Giustino and Filip obtained 25 perovskite compounds from 29 metals and 4 halide combinations and identified magnesium iodide perovskite as a potential candidate of which the bandgap is tunable in a range of visible light [46]. Körbel et al. obtained 199 (including novel 71) stable perovskite materials from over 32,000 possible ABX_3 compounds. Among the stable perovskite compounds, the only systems that are nontoxic and with a suitable electronic bandgap are tin and germanium halide perovskite [47]. Nakajima and Sawada proposed 51 low-toxic single- and double-metal halide perovskite materials from a materials library containing 11,025 compositions [48]. These material libraries will be expanded for datasets used in machine learning to generate materials showing novel properties.

1.6 Summary and Future Works

The power conversion efficiency of the perovskite solar cells has been rapidly increased to over 22% in the last decade. Process optimization, such as the two-step method and the antisolvent method for the crystallization of perovskite, strongly contributed to the rapid progress. However, why and how these processes contribute to effective photoenergy conversion is still unclear. Theoretical understanding of the processes is needed for suggestion of the indicator of process optimization to higher power conversion efficiency. Although the theoretical investigations summarized in this chapter were only small parts of the contribution for the developments, continuous theoretical studies strongly coupled with experimental evidence should be done toward further progress in this field.

References

1. Nakazaki, J., and Segawa, H. (2018). *J. Photochem. Photobiol. C*, **35**, 74–107.
2. Kojima, A., Teshima, K., Shirai, Y., and Miyasaka, T. (2009). *J. Am. Chem. Soc.*, **131**, 6050–6051.
3. Giorgi, G., Fujisawa, J.-I., Segawa, H., and Yamashita, K. (2013). *J. Phys. Chem. Lett.*, **4**, 4213–4216.
4. Dong, Q., Fang, Y., Shao, Y., Mulligan, P., Qiu, J., Cao, L., and Huang, J. (2015). *Science*, **347**, 967–970.
5. Tanaka, K., Takahashi, T., Ban, T., Kondo, T., Uchida, K., and Miura, N. (2003). *Solid State Commun.*, **127**, 619–623.
6. Jono, R., and Segawa, H. (2019). *Chem. Lett.*, **48**, 877–880.
7. Even, J., Pedesseau, L., Jancu, J.-M., and Katan, C. (2013). *J. Phys. Chem. Lett.*, **4**, 2999–3005.
8. Pedesseau, L., Jancu, J.-M., Rolland, A., Deleporte, E., Katan, C., and Even, J. (2014). *Opt. Quant. Electron.*, **10**, 1225–1232.

9. Even, J., Pedesseau, L., Jancu, J.-M., and Katan, C. (2014). *Phys. Status Solidi RRL*, **8**, 31–35.
10. Wang, Y., Gould, T., Dobson, J., Zhang, H., Yang, H., Yao, X., and Zhao, H. (2014). *Phys. Chem. Chem. Phys.*, **16**, 1424–1429.
11. Even, J., Pedesseau, L., and Katan, C. (2014). *Phys. Chem. Chem. Phys.*, **16**, 8697–8698.
12. Wang, Y., and Zhao, H. (2014). *Phys. Chem. Chem. Phys.*, **16**, 8699–8700.
13. Umari, P., Mosconi, E., and De Angelis, F. (2014). *Sci. Rep.*, **4**, 4467.
14. Brivio, F., Butler, K. T., Walsh, A., and van Schilfgaarde, M. (2014). *Phys. Rev. B*, **89**, 155204.
15. Even, J., Pedesseau, L., and Katan, C. (2014). *J. Phys. Chem. C*, **118**, 11566–11572.
16. Menéndez-Proupin, E., Palacios, P., Wahnón, P., and Conesa, J. C. (2014). *Phys. Rev. B*, **90**, 045207.
17. Geng, W., Zhang, L., Zhang, Y.-N., Lau, W.-M., and Liu, L.-M. (2014). *J. Phys. Chem. C*, **118**, 19565–19571.
18. Zhu, X., Su, H., Marcus, R. A., and Michel-Beyerle, M. E. (2014). *J. Phys. Chem. Lett.*, **5**, 3061–3065.
19. Feng, J., and Xiao, B. (2014). *J. Phys. Chem. C*, **118**, 19655–19660.
20. Buin, A., Pietsch, P., Xu, J., Voznyy, O., Ip, A. H., Comin, R., and Sargent, E. H. (2014). *Nano Lett.*, **14**, 6281–6286.
21. Filippetti, A., Delugas, P., and Mattoni, A. (2014). *J. Phys. Chem. C*, **118**, 24843–24853.
22. Brgoch, J., Lehner, A. J., Chabinyc, M., and Seshadri, R. (2014). *J. Phys. Chem. C*, **118**, 27721–27727.
23. Melissen, S. T. A. G., Labat, F., Sautet, P., and Le Bahers, T. (2015). *Phys. Chem. Chem. Phys.*, **17**, 2199–2209.
24. Mosconi, E., Umari, P., and De Angelis, F. (2015). *J. Mater. Chem. A*, **3**, 9208–9215.
25. Katan, C., Pedesseau, L., Kepenekian, M., Rolland, A., and Even, J. (2015). *J. Mater. Chem. A*, **3**, 9232–9240.
26. Etienne, T., Mosconi, E., and De Angelis, F. (2016). *J. Phys. Chem. Lett.*, **7**, 1638–1645.
27. Yu, Z.-G. (2016). *J. Phys. Chem. Lett.*, **7**, 3078–3083.
28. Azarhoosh, P., McKechnie, S., Frost, J. M., and Walsh, A. (2016). *APL Mater.*, **4**, 091501.
29. Yu, Z.-G. (2017). *Phys. Chem. Chem. Phys.*, **19**, 14907–14912.

30. Filip, M. R., Eperon, G. E., Snaith, H. J., and Giustino, G. (2014). *Nat. Commun.*, **5**, 5757.

31. Amat, A., Mosconi, E., Ronca, E., Quarti, C., Umari, P., Nazeeruddin, M. K., Grätzel, M., and De Angelis, F. (2014). *Nano Lett.*, **14**, 3608–3616.

32. Goldschmidt, V. M. (1926). Die Gesetze der Krystallochemie. *Naturwissenschaften*, **14**, 477–485.

33. Poglitsch, A., and Weber, D. (1987). *J. Chem. Phys.*, **87**, 6373–6378.

34. Ong, K. P., Goh, T. W., Xu, Q., and Huan, A. (2015). *J. Phys. Chem. Lett.*, **6**, 681–685.

35. Ong, K. P., Goh, T. W., Xu, Q., and Huan, A. (2015). *J. Phys. Chem. C*, **119**, 11033–11038.

36. Brivio, F., Walker, A. B., and Walsh, A. (2013). *APL Mater.*, **1**, 042111.

37. Motta, C., El-Mellouhi, F., Kais, S., Tabet, N., Alharbi, F., and Sanvito, S. (2015). *Nat. Commun.*, **6**, 7026.

38. Bechtel, J. S., Seshadri, R., and Van der Ven, A. (2016). *J. Phys. Chem. C*, **120**, 12403–12410.

39. Kanno, S., Imamura, Y., Saeki, A., and Hada, M. (2017). *J. Phys. Chem. C*, **121**, 14051–14059.

40. Chem, T., Foley, B. J., Park, C., Brown, C. M., Harriger, L. W., Lee, J., Ruff, J., Yoon, M., Choi, J. J., and Lee, S.-H. (2016). *Sci. Adv.*, **2**, e1601650.

41. Yin, W.-J., Shi, T., and Yan, Y. (2014). *Appl. Phys. Lett.*, **104**, 063903.

42. Shi, T., Yin, W.-J., and Yan, Y. (2014). *J. Phys. Chem. C*, **118**, 25350–25354.

43. Haruyama, J., Sodeyama, K., Han, L., and Tateyama, Y. (2014). *J. Phys. Chem. Lett.*, **5**, 2903–2909.

44. Uratani, H., and Yamashita, K. (2017). *J. Phys. Chem. Lett.*, **8**, 742–746.

45. Wang, W.-W., Dang, J.-S., Jono, R., Segawa, H., and Sugimoto, M. (2018). *Chem. Sci.*, **9**, 3341–3353.

46. Filip, M. R., and Giustino, F. (2016). *J. Phys. Chem. C*, **120**, 166–173.

47. Körbel, S., Marques, M. A. L., and Botti, S. (2016). *J. Mater. Chem. C*, **4**, 3157–3167.

48. Nakajima, T., and Sawada, K. (2017). *J. Phys. Chem. Lett.*, **8**, 4826–4831.

Chapter 2

Electronic Properties of Organic–Inorganic Lead Halide Perovskite

Seongrok Seo, Seonghwa Jeong, and Hyunjung Shin
Department of Energy Science, Sungkyunkwan University, Suwon, Korea
hshin@skku.edu

2.1 Introduction

Organic–inorganic hybrid metal halides are now emerging as a new class of semiconducting materials. Known as "perovskite solar cell," methylammonium lead triiodide ($MAPbI_3$), one of the notable examples of hybrid metal halides, shows a remarkable power conversion efficiency of over 20% as a light-absorbing materials [1–11].

$MAPbI_3$ has a crystal structure of perovskite with the general chemical formula ABX_3 in a simple cubic lattice [12]. Here, A is an organic methylammonium (CH_3NH_3) ion (MA^+), B is typically a Pb ion (Pb^{2+}), and X is an iodine ion (I^-). Pb and iodine ions construct a PbI_6 octahedra as a framework. It can be viewed as an anion corner-shared 3D network of a PbI_6 octahedra, as shown in Fig. 2.1 [13]. As a simple cubic lattice the Pb ions are centered in the octahedra and

Multifunctional Organic–Inorganic Halide Perovskite: Applications in Solar Cells, Light-Emitting Diodes, and Resistive Memory
Edited by Nam-Gyu Park and Hiroshi Segawa

ISBN 978-981-4800-52-5 (Hardcover), 978-1-003-27593-0 (eBook)
www.jennystanford.com

the MA ions are sitting in between the octahedra [14]. The highest symmetry phase for $MAPbI_3$ and related materials is the cubic (Pm3m) lattice, with sequential transitions lowering the symmetry typically through octahedral tilting. The tilting collectively leads to the phase transition from cubic to tetragonal (Fig. 2.2) [15]. A collective rotation of the PbI_6 octahedra around the *c* axis results in a tetragonal space group I_4/mcm, with closer packing within the *ab* plane. A further transition to an orthorhombic phase (space group of Pnma) is accompanied by a tilting of the PbI_6 octahedra out of the *ab* plane. $MAPbI_3$ transforms from cubic to tetragonal at temperatures above 315 K $\approx$ 330 K and tetragonal to orthorhombic phase at ~160 K. These transitions need to be critically considered when assessing charge carrier dynamics because they may alter the electronic band structure and therefore optoelectronic properties of the material.

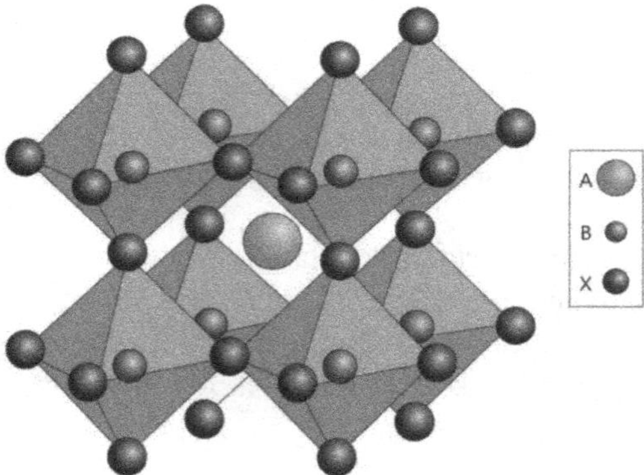

Figure 2.1 Perovskite crystal structure. For photovoltaically interesting perovskites, the large cation A is usually the methylammonium ion (CH_3NH_3), the small cation B is Pb, and the anion X is a halogen ion (usually I, but both Cl and Br are also of interest). For $CH_3NH_3PbI_3$, the cubic phase forms only at temperatures above 330 K. Reprinted by permission from Macmillan Publishers Ltd: *Nature Photonics* (Ref. [13]), copyright (2014).

In this chapter, fundamental electronic properties of $MAPbI_3$ and related materials, including crystal structure, electronic band structure, direct-indirect bandgap character, strong spin-orbit coupling, effective masses of electrons and holes, optical absorption, defects and their tolerance, and photoexcited charge carrier transport property, are described. As the transport properties related with their electronic structures, slow recombination rate, direct-indirect nature, and their mobility will be summarzed.

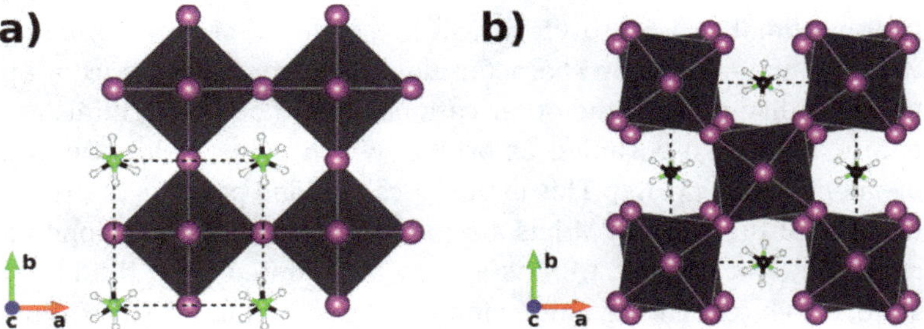

Figure 2.2 Schematic structural transformation representation of $MAPbI_3$, where MA = methylammonium, in the (a) pseudocubic and (b) tetragonal phase. Thick dashed lines indicate the unit cell, which contains carbon (black), nitrogen (green), hydrogen (white), iodine (violet), and lead (light-gray) atoms; the latter are inside of the dark-shaded octahedra. For convenient visualization, atoms belonging to more than a standard unit cell are shown. Reprinted with permission from Ref. [15]. Copyright (2014) American Chemical Society.

2.2 Electronic Band Structure

Understanding the origin of the electronic structure of $MAPbI_3$ material as an absorber is of importance. The maximum attainable solar cell efficiency is directly dependent upon the absorber bandgap. Furthermore, a qualitatively sound picture of the electronic bandgap may also provide insight into the transport behavior of photoexcited charge carriers in such materials. As a result, photovoltaic devices can be optimized [16, 17].

2.2.1 Electronic Bandgap

Many theoretical investigations have already contributed to our fundamental electronic structural understanding [18, 19]. In contrast, demonstration of the fundamental experimental findings is still awaited [20]. In 2014, the electronic structures of $TiO_2/MAPbI_3$ interfaces were investigated using photoelectron spectroscopy with hard X-ray. Simulated density of states (DOS) of the valence structure describes the experimental spectra well enough and shows that the outermost levels consist of lead and iodine orbitals [21]. For

$MAPbI_3$ and related materials, the valence band maximum (VBM) is predominantly composed of I 5p orbitals, while the conduction band minimum (CBM) is formed of Pb 6p. More detailed, the VBM is approximately I 5p to Pb 6s (3:1), and the CBM is nonbonding with the majority Pb 6p character along with contribution from I p states. Unlike most of the other cations, whose outer s orbitals are empty, Pb has an occupied 6s orbital, which is just below the top of valence bands (VBs). This is the so-called lone pair of s electrons in Pb. And thus, the VBM has a strong Pb 6s and I 5p antibonding character, whereas the CBM is almost contributed from the Pb 6p state, which reflects the unique ionic and covalent nature of $MAPbI_3$ [22, 23]. Specifically, the electronic structure near the band edge in $MAPbI_3$ is primarily dictated by the basic PbI_6 building blocks. The bonding diagram of isolated $[PbI_6]^{4-}$ provides a foundation for understanding more complex band structures, as shown in Fig. 2.3a. For the isolated $[PbI_6]^{4-}$ units, a Pb 6s – I 5p σ-antibonding orbital comprises the highest occupied molecular orbital, while Pb 6p – I 5p π-antibonding and Pb 6p – I 5s s-antibonding orbitals constitute the lowest unoccupied molecular orbital. Figure 2.3b is the electronic band structure of $MAPbI_3$. Orbital contributions are indicated by I 5p (green), Pb 6p (red), and Pb 6s (blue) [24–27].

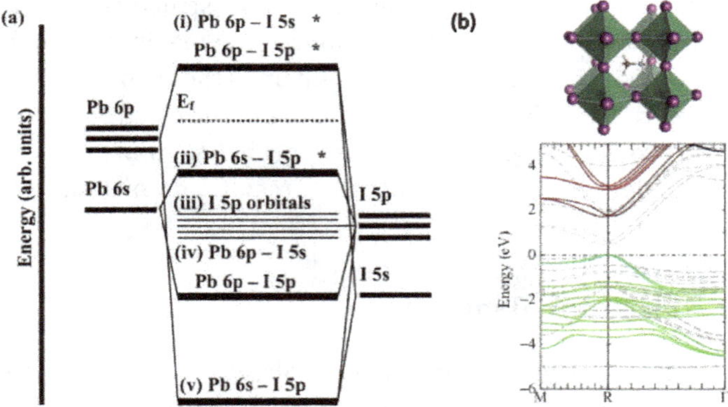

Figure 2.3 (A) Bonding diagram of an isolated $[PbI_6]^{4-}$ octahedra without considering the SOC. Reprinted figure with permission from Ref. [24]. Copyright (2003) by the American Physical Society. (B) Calculated electronic band structure of $CH_3NH_3PbI_3$. Orbital contributions are indicated by I 5p (green), Pb 6p (red), and Pb 6s (blue). Dashed gray lines show results from LDA calculations, which severely underestimate E_g and dispersion at near VBM. Reprinted with permission from Ref. [25]. Copyright (2014) by the American Physical Society.

In conventional semiconductors such as GaAs and CdTe, the CBM primarily has an s orbital character whereas the VBM has a p orbital character. As discussed, $MAPbI_3$ exhibits an inverted band structure. And also the MA^+ ion contribution to the states is far from band edges. That is, occupied molecular states are found well below the top of the VB and empty molecular states are found well above the bottom of the conduction bands (CBs). As a consequence, the organic molecules do not play a direct role in forming the electronic structures, rather than stabilizing the perovskite's structure and changing the lattice constant [28, 29]. As a result, $MAPbI_3$ and related materials are direct-bandgap semiconductors with strong band optical absorption and luminescence. That is, the VBM and CBM lie at the same point in reciprocal space (k-space).

Note that recent theoretical calculations suggest that there may be subtle yet important exceptions that arise in noncentrosymmetric crystal structures in $MAPbI_3$ due to spin-orbit coupling (SOC) [30, 31]. The Rashba spin-splittings with VBM and CBM are not located at the same high-symmetry points of the Brillouin zone (R point) but are instead slightly shifted. In other words, the CBM and VBM are not located at the same points, having different Rashba momentum offsets [32–35]. This result indicates the formation of an indirect bandgap, which significantly reduces the radiative recombination rate. On the other hand, the absorption spectrum is not largely affected by the presence of an indirect gap like this simply because the indirect bandgap is generally only a few tens of meV smaller than the direct optical transitions gap [36]. Furthermore, direct bandgap characters in strong optical absorption and indirect characters in a slow recombination rate are responsible for the high efficiencies of $MAPbI_3$ solar cells.

Both Pb and I are heavy ions in $MAPbI_3$, and as such both valence and conduction bands contain considerable relativistic effects, that is, for a quantitative treatment of the electronic band structure SOC must be included. The SOC results in a decrease in the electronic bandgap, which can be attributed to the splitting of the gap. The bandgap predicted by a state-of-the-art quasiparticle self-consistent GW with SOC corrections is the only approach to predict accurate electronic structure of $MAPbI_3$. Calculated direct bandgap of $MAPbI_3$ E_g is ~1.67 eV, which is in good agreement with the measured value of 1.61 eV from room temperature photoluminescence. The computed bulk ionization potential and electron affinity are 5.7 eV and 4.1 eV, respectively [37].

2.2.2 Effective Masses of Electrons and Holes

Another interesting feature of the band structure is the comparable-at-the-same-time dispersive parabolicity of the upper VB and lower CB near the R point in *k*-space. According to the well-known parabolic approximation for effective mass as derived from band dispersion, small and at the same time similar dispersion at the band edge of $MAPbI_3$ results in light electrons as well as holes with comparable effective masses. Intuitively, the more dispersive (broader) the band near the band edges, the lighter (heavier) is the effective mass. In conventional thin film solar cell absorbers, such as GaAs and CdTe, the CBM is mostly contributed by anions' p character, the so-called p-s semiconductor. The s orbitals in the higher energy level are more delocalized than the p orbitals and thus, the lowest CB is more dispersive than the highest VB. Consequently, the electron's effective mass is much smaller than the hole's. This is why electrons are transported much faster than holes in conventional semiconductors.

As noted earlier, the electronic structure of $MAPbI_3$ is inverted compared to conventional p-s semiconductors. Its CBM is derived from Pb p orbitals, and the VBM is a mixture of Pb s and I p (s-p semiconductor) orbitals. The lower CB of $MAPbI_3$ is more dispersive than the upper VB in p-s semiconductors. On the other hand, due to strong s-p coupling around the VBM, the upper VB of $MAPbI_3$ is also dispersive. Recent calculations using SOC-GW provided average values of $m_e/m_0 = 0.19$ and $m_h/m_0 = 0.25$ for $MAPbI_3$, where m_0 is the free electron mass. These showed good agreement with effective masses determined recently using high-field magnetoabsorption spectroscopy [38]. Considering mobility is inversely propositional to carrier effective mass, it is evident that $MAPbI_3$ shows rather balanced, or ambipolar, electron and hole diffusion lengths [39]. As a result, the strong s-p antibonding coupling in $MAPbI_3$ leads to small effective masses of electrons and holes and thus even more efficient solar cells in the hole transporting layer/$MAPbI_3$/electron transporting layer configuration.

2.2.3 Optical Absorption

In a semiconductor, the optical absorption at a given frequency is a function of two primary factors: (i) the joint DOS and (ii) the transition matrix from states in the VB to states in the CB. The former measures the total number of possible photoelectron transitions, and the latter measures the probability of each photoelectron transition. And thus, the optical absorption coefficient of a material is closely related to its electronic band structure.

An important feature of $MAPbI_3$ and related materials is their relatively large absorption coefficient. $MAPbI_3$ materials, for instance, less than 500 nm in thickness can absorb enough photons to achieve a high efficiency of well above 15%. In comparison, the thicknesses of the absorber layers in first- and second-generation solar cells are about 300 μm and 2 μm, respectively. For the first-generation solar cells mainly with Si, the absorption close to the band edge is from the Si p orbital to the Si p and s orbitals of the typical s-p semiconductor. However, Si is an indirect-bandgap material and its transition probability is 2 orders of magnitude lower than that of the direct-bandgap materials. Si solar cells usually require an absorbing layer 2 orders of magnitude thicker. GaAs as an example and $MAPbI_3$ are direct-bandgap semiconducting materials and thus their optical absorption is much stronger than that of Si. However, the electronic structure of $MAPbI_3$ is an inverted one compared with GaAs. The atomic p orbital shows less dispersion than s orbitals do. Figure 2.4 shows the schematic optical absorption of (a) first-generation, (b) second-generation, and (c) halide perovskite solar cell absorbers [40]. As a result, the DOS in the lower CB of the $MAPbI_3$ materials is significantly higher than that of GaAs, leading to a higher joint DOS, as shown in Fig. 2.5a and 2.5b [41]. The transition probability between Pb s and Pb p of $MAPbI_3$ is also high, which is the transition probability comparable to that of GaAs. Therefore, $MAPbI_3$ and related materials show stronger optical absorption than GaAs. The optical absorption coefficient of $MAPbI_3$ is up to 1 order of magnitude higher than that of GaAs within the visible range. In the literature, the optical absorption coefficient of $MAPbI_3$ is reported in the range of ~6×10^3 to ~4.5×10^4 cm^{-1}.

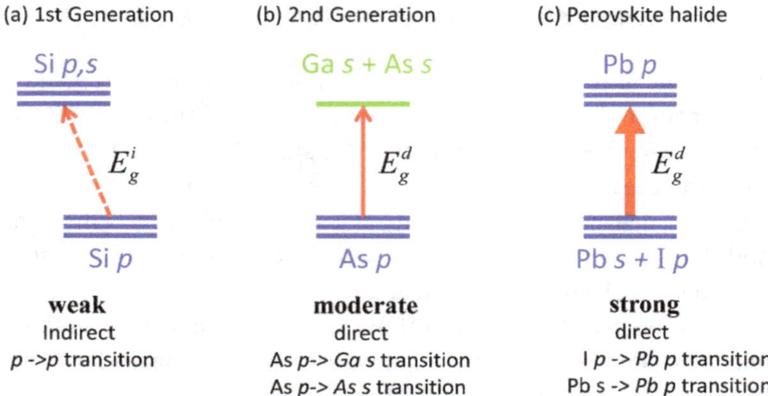

Figure 2.4 The schematic optical absorption of (a) first-generation, (b) second-generation, and (c) halide perovskite solar cell absorber. GaAs has been chosen as a prototypical second-generation solar cell absorbers. Reproduced from Ref. [40] with permission of The Royal Society of Chemistry.

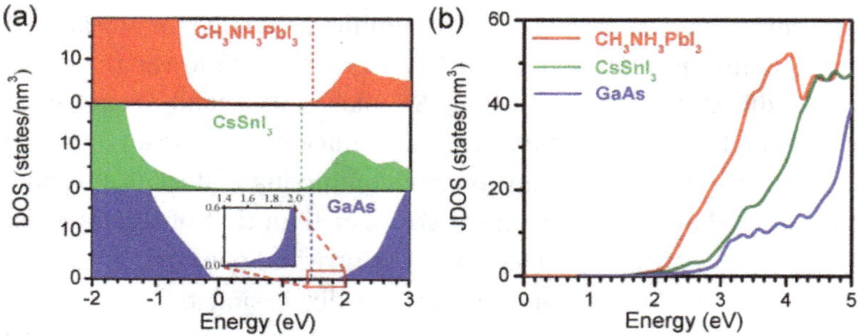

Figure 2.5 (A) Comparison of the DOS of $CH_3NH_3PbI_3$, $CsSnI_3$, and GaAs. (B) Calculated JDOS of the compounds in (A). Reprinted from Ref. [41], Copyright 2014, with permission of John Wiley & Sons.

2.2.4 Defects in a Bandgap

$MAPbI_3$ and related materials mainly synthesized a low-temperature solution process until now. The low-temperature solution process inevitably produces many different types of defects, including thermodynamically stable point defects, interfaces, surfaces, and grain boundaries [42]. Nonradiative recombination and carrier scattering are often caused by defects that generate deep level states in bandgap. Most of the structural defects, for example, surfaces and grain boundaries, are in deep level states [43]. Unusually, deep level defects with a high formation energy are scarce, whereas many shallow trap states in bandgaps can be found in $MAPbI_3$, resulting in defect tolerating materials. These unique defect properties are also attributed to the strong Pb lone pair s–halogen p antibonding coupling and the ionic characteristics [44–46].

First-principles calculations have been used to study the point defect properties of $MAPbI_3$ and related materials. The chemical potentials for equilibrium growth are shown in Fig. 2.6 [47, 48]. The growth of $MAPbI_3$ phase can occur only in a long and narrow range of chemical potential, indicating a small experimental window in terms of thermodynamic variables, for example, precursors, pressure, and temperature. All the possible intrinsic point defects are three vacancies (V_{MA}, V_{Pb}, V_I), three interstitials (MA_i, Pb_i, I_i), and four antisite substitutions (MA_I, Pb_I, I_{MA}, I_{Pb}). The formation energies of these point defects as a function of Fermi energy at chemical potentials A, B, and C are shown in Fig. 2.7. Note that all vacancy defects and most interstitial defects, such as MA_i, V_{Pb}, MA_{Pb}, I_i, V_I, and V_{MA}, exhibit rather shallow transition energy levels. The shallow level defects of both donor-like and acceptor-like have comparable low formation energies, resulting in both *n*- and *p*-type electronic nature in $MAPbI_3$. In $MAPbI_3$, the dominant donors MA_i and V_I and acceptors V_{Pb} and MA_{Pb} have comparable formation energies, so both *p*- and *n*-type doing are possible. The shallow acceptors originate from covalent coupling between the Pb lone-pair s and I p orbitals, which increases the VBM so that the acceptors are generally shallower than in cases without strong s-p coupling. In contrast, the shallow donors are due to the high ionicity. The MA_i has no covalent bonds to the Pb-I framework and does not create additional gap states. The calculated transition energy levels of point defects in $MAPbI_3$ are shown in Fig. 2.8. Theoretical bipolar conductivities have been already confirmed by doping experiments [49]. The defects that generate deep levels are I_{MA}, I_{Pb}, Pb_i, MA_I, and Pb_I; and mostly cation or antisite defects, except Pb_i, having high formation energies. Recently, the group of Sargent provided a complete electronic band diagram of single-crystal $MAPbI_3$ by an in-gap electronic state spectrum [50]. The mobility and diffusion lengths of both electrons and holes are directy measured, as well as the concentration and type of charge carriers [50].

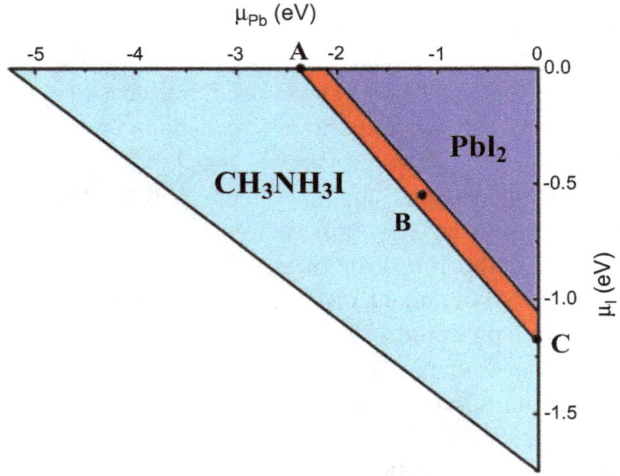

Figure 2.6 The thermodynamic stable range for equilibrium growth of $CH_3NH_3PbI_3$ is the narrow but long region marked in red color. Outside this region, the compound will decompose into PbI_2 and CH_3NH_3I. We choose representative points A($\mu_{MA} = -2.87$ eV, $\mu_{Pb} = -2.39$ eV, $\mu_I = 0$ eV), B($\mu_{MA} = -2.41$ eV, $\mu_{Pb} = -1.06$ eV, $\mu_I = -0.60$ eV), and C($\mu_{MA} = -1.68$ eV, $\mu_{Pb} = 0$ eV, $\mu_I = -1.19$ eV) for the formation energy study. Reprinted from Ref. [47] with permission from AIP Publishing.

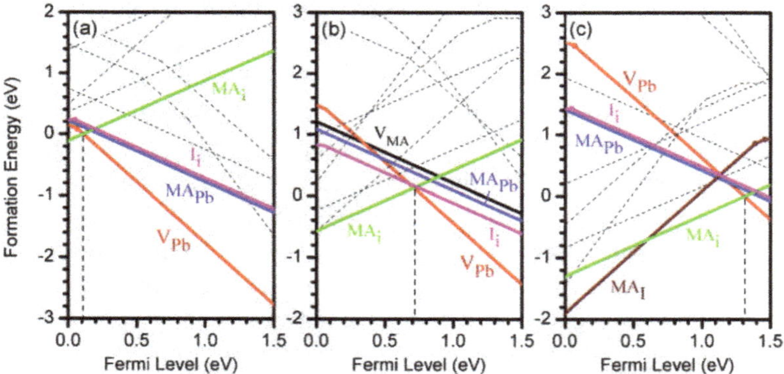

Figure 2.7 The formation energies of intrinsic point defects in $CH_3NH_3PbI_3$ at chemical potentials A, B, and C shown in Fig. 2.6. Defects with much higher formation energies have been displayed as dashed lines. Reprinted from Ref. [47] with permission from AIP Publishing.

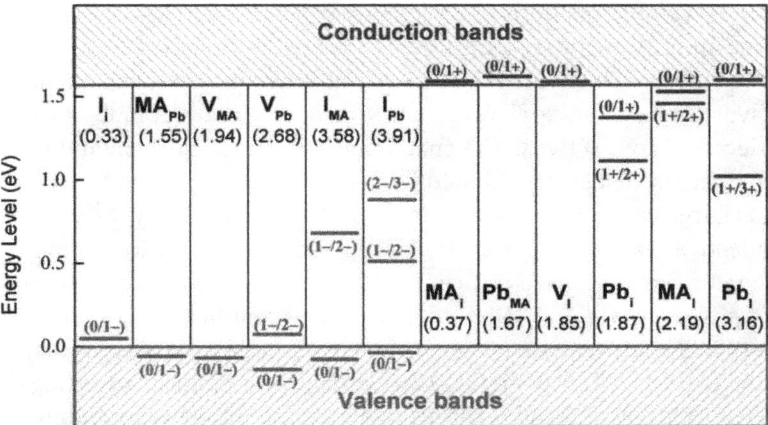

Figure 2.8 Calculated transition energy levels of point defects in $MAPbI_3$. The formation energies of neutral defects are shown in parentheses. The acceptors/donors are ordered by the formation energies (from left to right). MA denotes CH_3NH_3. Reprinted from Ref. [41], Copyright 2014, with permission of John Wiley & Sons.

2.2.5 Defect Tolerance

Conventional semiconductor technology is centered on producing perfect crystals to reduce defects, as exemplified by single-crystalline silicon with remarkably low defect densities. In $MAPbI_3$, however, synthesis by solution processing at near room temperature will inevitably result in a high density of defects that may be detrimental to its performance. In this context, the defect tolerating $MAPbI_3$ materials are still not fully understood.

The original defect tolerance concept stated that the presence of an antibonding upper VB and a bonding lower CB, like in $MAPbI_3$, signifies that dangling bond defects would be repelled quantum-mechanically into the continuum bands, leaving the bandgap clean and allowing for the formation of shallow defects. The second point to consider is a screening in the dielectric medium. If charged defects are present in the host material, their influence should be minimized. The dielectric constant represents the ability of a material to screen an electrostatic perturbation. The higher the dielectric constant of a material, the more effective is the screening, and thus the materials show better defect tolerance. The dielectric constants found in

$MAPbI_3$ are roughly three times larger than those found in other thin film PV materials, such as CdTe and Cu_2ZnSnS_4 [51, 52]. The last factor to consider will be carrier effective masses. A small mass favors free charge carriers and can support high carrier mobility and electrical conductivity. The spatial localization of electron and hole at charged defect sites should be avoided as it slows the transport of charge carriers and is associated with thermal energy losses. The effective mass is also a critical factor in avoiding the formation of small polarons. There is a competition between the kinetic energy of a free carrier and the potential energy gain by localizing in the lattice. Typically, metal oxides often have a large hole effective mass due to the localization of the O 2p orbitals forming the VB, leading to the formation of small polarons and thus the difficulty in finding of p-type oxides.

In $MAPbI_3$, the unique electronic structure results from the framework of the corner-sharing PbI_6^{3-} octahedral with overall stoichiometry PbI^{3-} and a sublattice of MA^+ cations in the cuboctaheral voids, as discussed earlier. The framework structure is intrinsically soft and shows dynamical disorder. The relatively light electrons and/or holes in the CB are rapidly protected to form a large polaron with phonons [53]. And also the large polaron can screen electrostatic potential drastically and thus reduces its scattering with charged defect sites [54]. As a result, the high polarizability in the soft and ionic lattice can lead to the sufficient screening and compensation of charged defects in the $MAPbI_3$ lattice. The traps are energetically shallow and electrostatically screened, thus diminishing the role of these traps in recombination processes [55, 56].

2.3 Photogenerated Charge Carrier Transport

The remarkable charge transport behaviors of $MAPbI_3$ [57, 58], such as large diffusion length [59], low recombination rate, high mobility, and long carrier lifetime seem to suggest that charge carriers originate from scattering with defects, longitudinal optical phonons, and other carriers [60, 61]. Even though a lot of focus is on theoretical calculations and experimental findings, the long charge carrier lifetime and mobility mechanism is not clear yet.

2.3.1 Slow Recombination Rate

The outstanding photovoltaic performance of $MAPbI_3$ and related materials is mainly due to the high absorption coefficient and low exciton binding energy [62, 63] and thus the high yield of free electrons and holes following photoexcitation and excellent charge transport properties [64, 65]. So far, $MAPbI_3$ is believed to be a direct-bandgap semiconductor, where the absorption and emission of photons occur via allowed transitions. In Si as an indirect-bandgap semiconductor, the absorption and recombination involve not only photons but also phonons. This results in lower absorption coefficients than in direct semiconductors, but at the same time recombination is much slower. As discussed earlier, recent theoretical calculations with a strong Pb SOC in $MAPbI_3$ show that the CBM is slightly shifted in k-space with respect to the VBM, making the fundamental bandgap indirect. The group of Stranks and Savenije [66] proposed that the bandgap in $MAPbI_3$ has a direct-indirect character, as shown in Fig. 2.9. They showed that second-order electron-hole recombination of photoexcited charges is retarded at a lower temperature by time-resolved photoconductance measurements. According to their experimental results, the indirect bandgap is to lie at 47 ≈ 75 meV below the direct bandgap in $MAPbI_3$. Their observation is consistent with a slow phonon-assisted recombination pathway via the indirect bandgap. Numerical simulations also addressed the disparate roles of direct and indirect bandgaps in absorption and emission. They are found to be general for solution-processed $MAPbI_3$ and related materials in the tetragonal phase and are compatible with theoretical predictions for the mixed direct-indirect bandgap character. In their analysis, electrons photoexcited in the CB thermally relax into dark states in band-like structures, yielding highly mobile carriers. Fast radiative recombination of these electrons with VB holes is forbidden, thus resulting in an extended charge carrier lifetime. This implies that the radiative electron-hole recombination is momentum-forbidden, as in the case of an indirect-bandgap semiconductor.

The general dynamics of charge carrier recombination [67] through monomolecular and higher-order processes are governed by the following rate equation:

$$\frac{dn}{dt} = G - k_1 n - k_2 n^2 - k_3 n^3 = G - nR_T(n), \tag{2.1}$$

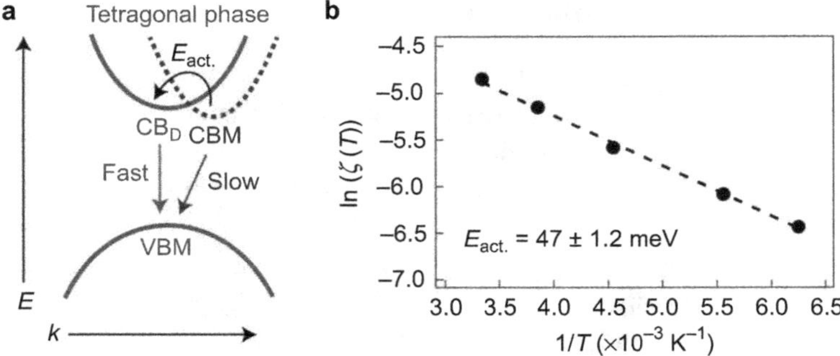

Figure 2.9 Proposed band diagram and activation energy for second-order recombination in the $MAPbI_3$ thin film. (a) Proposed band diagram for the tetragonal phase. Here, the conduction band minimum (CBM) is slightly shifted in *k*-space with respect to the valence band maximum (VBM), making the fundamental bandgap indirect. The local minimum in CB corresponding to a direct transition is denoted by CB_D. (b) Arrhenius plot of the prefactor $\zeta(T)$, defined as the ratio between the $k_2(T)$, obtained from fitting the experimental TRMC traces for 160 K < *T* < 300 K, and the theoretical second-order recombination rate $B(T)$. The slope of the linear fit (dashed line) indicates that the activation energy for second-order recombination is 47 ± 1.2 meV. Therefore, thermal energy (k_BT) might assist in the release of electrons from CBM to CB_D, which is schematically depicted in (a). Reprinted by permission from Macmillan Publishers Ltd: *Nature Materials* (Ref. [66]), copyright (2014).

where G is the charge-density generation rate, k_1 the monomolecular charge-recombination rate constant, k_2 the bimolecular electron-hole recombination rate constant, and k_3 the Auger recombination rate constant. Here $R_T(n)$ is the total charge recombination rate, given by

$$R_T = k_1 + nk_2 + n^2k_3, \tag{2.2}$$

which depends on the charge carrier density, n, and the pulsed excitation time, t. Monomolecular charge carrier recombination is by definition a process involving one "particle," which is a CB electron, a VB hole, or an exciton composed of a bound electron-hole pair. Due to the quite low exciton binding energy in $MAPbI_3$ [68], the predominant species present at room temperature are free charge carriers rather than excitons. Only one characteristic recombination kinetic was observed independent of the measurement methods. Therefore, the monomolecular decay component observed most likely originated from trap-assisted recombination. The trap-mediated recombination rate k_1 for $MAPbI_3$ over the temperature range of 8–340 K was found to decrease monotonically with a decreasing temperature. From an Arrhenius analysis, the activation energy for such trap passivation was about 20 meV in the room temperature tetragonal phase of $MAPbI_3$ films, suggesting the presence of predominantly shallow traps.

Bimolecular charge carrier recombination between electrons and holes in a direct semiconductor can be simply viewed as intrinsic photon-radiative recombination [69]. The rate constants, k_2, were measured in the range of 0.6×10^{-10} and 14×10^{-10} $cm^3\ s^{-1}$ at room temperature. It is surprisingly comparable with the k_2 of $\sim 4 \times 10^{-10}$ $cm^3\ s^{-1}$ for the direct single-crystalline inorganic semiconductor GaAs. This can be understood by the indirect band structure of $MAPbI_3$ with strong SOC, as explained.

2.3.2 Charge Carrier Mobility [70]

The Langevin model assumes that recombination will occur once an electron and a hole move within their joint capture radius, which is presumed to be larger than their mean free path. The model is usually utilized for materials with relatively low charge carrier mobility, whereas it fails to explain the charge carrier mobility in $MAPbI_3$. Non-Langevin recombination behavior was attributed to a spatial charge separation between electrons and holes so that similar effects may arise in $MAPbI_3$. Therefore, weak preferential localization of electrons and holes in different regions may cause a reduction in the spatial overlap and hence recombination rates [71, 72]. It turns out to be highly successful in modeling the radiative recombination in high-charge carrier mobility semiconducting materials, such as GaAs and $MAPbI_3$ [73, 74].

A polaron is an excess electron or hole dressed by polarization of nuclear coordinates in a crystalline lattice and is ubiquitous to polar and/or polarizable solids. The electron-phonon coupling may be strong enough to result in a self-trapped polaron, that is, the electron (hole) dressed by the nuclear polarization. For $MAPbI_3$, charge formation and transport through the soft and ionic lattice with crystal-liquid duality has to be critically considered [75–77]. Given the exceptionally high polarizable soft lattice, a polaron must form for an excess charge. And thus, a large polaron is delocalized over a few unit cells and its transport is coherent and band-like, with carrier mobility (μ) decreasing with an increasing temperature (T), that is, $d\mu/dT < 0$. In contrast, a small polaron is localized to a unit cell and its transport occurs via thermally activated hopping, that is, $d\mu/dT > 0$. In this regard, large polaron transport resembles the coherent transport of a free electron (hole) in the CB. Compared to a free band electron, a fascinating consequence of the heavier mass of a large polaron is the much-reduced scattering with phonons or defects. The polarons in $MAPbI_3$ must be large polarons, because both transport and spectroscopic measurements showed $d\mu/dT < 0$ in a broad temperature range. In both tetragonal and cubic phases, the observed temperature dependencies ($d\mu/dT < 0$) establish coherent transport but different scaling laws, $\mu \ \alpha \ T^{-1.5}$ and $T^{-0.5}$, respectively. It may suggest the dominance of different scattering mechanisms in cubic and tetragonal phases. As a consequence, the large polaron formation in $MAPbI_3$ is the efficient screening of the Coulomb potential showing the exceptional defect tolerance [78–81].

2.4 Conclusion

$MAPbI_3$ and related materials show superior performances as absorbers in photovoltaic cells. Since they are a novel class of semiconducting material, their electronic and optoelectronic properties are of importance to understand the high photon energy conversion efficiency. Based on the fundamental electronic band structure of $MAPbI_3$, strong absorption, effective masses of electrons and holes, slow recombination rate, direct-indirect bandgap behavior with a strong SOC, defect tolerance, and finally the photogenerated electrons/holes forming the large polarons are critical processes and/or mechanisms attributing to highly efficient solar cells.

References

1. Kojima, A., Teshima, K., Shirai, Y., and Miyasaka, T. (2009). Organometallic halide perovskite as visible-light sensitizers for photovoltaic cells. *J. Am. Chem. Soc.*, **131**, 6050–6051.

2. Im, J.-H., Lee, C.-R., Lee, J.-W., Park, S.-W., and Park, N.-G. (2011). 6.5% efficient perovskite quantum-dot-sensitized solar cell. *Nanoscale*, **3**, 4088.
3. Kim, H.-S., Lee, C.-R., Im, J.-H., Lee, K.-B., Moehl, T., Marchioro, A., Moon, S.-J., Humphry-Baker, R., Yum, J. H., Moser, J. E., Gratzel, M., and Park, N.-G. (2009). Lead iodide perovskite sensitized all-solid-state submicron thin film mesoscopic solar cell with efficiency exceeding 9%. *Sci. Rep.*, **2**, 591.
4. Etgar, L., Gao, P., Xue, A., Peng, Q., Chandiran, A. K., Liu, B., Nazeeruddin, Md. K., and Gratzel, M. (2012). Mesoscopic $CH_3NH_3PbI_3/TiO_2$ heterojunction solar cells. *J. Am. Chem. Soc.*, **134**, 17396–17399.
5. Lee, M. M., Teuscher, J., Miyasaka, T., Murakami, T. N., and Snaith, H. (2012). Efficient hybrid solar cells based on meso-superstructured organometal halide perovskites. *Science*, **338**, 643–647.
6. Snaith, H. J. (2013). Perovskite: the emergence of a new era for low-cost, high-efficiency solar cells. *J. Phys. Chem. Lett.*, **4**, 3623–3630.
7. Lin, Q., Armin, A., Burn, P. L., and Meredith, P. (2016). Organohalide perovskites for solar energy conversion. *Acc. Chem. Res.*, **49**, 545–553.
8. Correa-Baena, J.-P., Abate, A., Saliba, M., Tress, W., Jacobsson, T. J., Gratzel, M., and Hagfeldt, A. (2017). The rapid evolution of highly efficient perovskite solar cells. *Energy Environ. Sci.*, **10**, 710–727.
9. Green, M. A., and Ho-Baillie, A. (2017). Perovskite solar cells: the birth of a new era in photovoltaics. *ACS Energy Lett.*, **2**, 822–830.
10. Manser, J. S., Christians, J. A., and Kamat, P. V. (2016). Intriguing optoelectronic properties of metal halide perovskites. *Chem. Rev.*, **116**, 12956–13008.
11. Zhao, Y., and Zhu, K. (2016). Organic-inorganic hybrid lead halide perovskites for optoelectronic and electronic applications. *Chem. Soc. Rev.*, **46**, 655–689.
12. Baikie, T., Fang, Y., Kadro, J. M., Schreyer, M., Wei, F., Mhaisalkar, S. G., Gratzel, M., and White, T. J. (2013). Synthesis and crystal chemistry of the hybrid perovskite $(CH_3NH_3)PbI_3$ for solid-state sensitised solar cell applications. *J. Mater. Chem. A*, **1**, 5628.
13. Green, M. A., Ho-Baillie, A., and Snaith, H. J. (2014). The emergence of perovskite solar cells. *Nat. Photonics*, **8**, 506–514.
14. She, L., Liu, M., and Zhong, D. (2016). Atomic structures of $CH_3NH_3PbI_3$ (001) surfaces. *ACS Nano*, **10**, 1128–1131.
15. Egger, D. A., and Kronik, L. (2013). Role of dispersive interactions in determining structural properties of organic-inorganic halide

perovskites: insights from first-principles calculations. *J. Phys. Chem. Lett.*, **5**, 2728–2733.

16. Frost, J. M., Butler, K. T., Brivio, F., Hendon, C. H., van Schilfgaarde, M., and Walsh, A. (2014). Atomistic origins of high-performance in hybrid halide perovskite solar cells. *Nano Lett.*, **14**, 2584–2590.
17. Yang, D., Lv, J., Zhao, X., Xu, Q., Fu, Y., Zhan, Y., Zunger, A., and Zhang, L. (2017). Functionality-directed screening of Pb-free hybrid organic-inorganic perovskites with desired intrinsic photovoltaic functionalities. *Chem. Mater.*, **29**, 524–538.
18. Mosconi, E., Amat, A., Nazeeruddin, M. K., Gratzel, M., and De Angelis, F. (2013). First-principles modeling of mixed halide organometal perovskites for photovoltaic applications. *J. Phys. Chem. C*, **117**, 13902–13913.
19. Yun, S., Zhou, X., Even, J., and Hagfeldt, A. (2017). Theoretical treatment of $CH_3NH_3PbI_3$ perovskite solar cells. *Angew. Chem. Int. Ed.*, **56**, 2–14.
20. Hoye, R. L. Z., et. al. (2017). Perovskite-inspired photovoltaic materials: toward best practices in materials characterization and calculations. *Chem. Mater.*, **29**, 1964–1988.
21. Lindblad, R., Bi, D., Park, B.-W., Oscarsson, J., Gorgoi, M., Siegbahn, H., Odelius, M., Johansson, E. M. J., and Rensmo, H. (2014). Electronic structure of $TiO_2/CH_3NH_3PbI_3$ perovskite solar cell interfaces. *J. Phys. Chem. Lett.*, **5**, 648–653.
22. Buin, A., Pietsch, P., Xu, J., Voznyy, O., Ip, A. H., Comin, R., and Sargent, E. H. (2014) Materials processing routes to trap-free halide perovskites. *Nano Lett.*, **14**, 6281–6286.
23. Emara, J., Schnier, T., Pourdavoud, N., Riedl, T., Meerholz, K., and Olthof, S. (2016). Impact of film stoichiometry on the ionization energy and electronic structure of $CH_3NH_3PbI_3$ perovskites. *Adv. Mater.*, **28**, 553–559.
24. Umebayashi, T., and Asai, K. (2003). Electronic structures of lead iodide based low-dimensional crystals. *Phys. Rev. B*, **67**, 155405.
25. Brivio, F., Butler, K. T., and Walsh, A. (2014). Relativistic quasiparticle self-consistent electronic structure of hybrid halide perovskite photovoltaic absorbers. *Phys. Rev. B*, **89**, 155204.
26. Endres, J., Egger, D. A., Kulbak, M., Kerner, R. A., Zhao, L., Silver, S. H., Hodes, G., Rand, B. P., Cahen, D., Kronik, L., and Kahn, A. (2016). Valence and conduction band densities of states of metal halide perovskites: a combined experimental-theoretical study. *J. Phys. Chem. Lett.*, **7**, 2722–2729.

27. Liu, G., Kong, L., Gong, J., Uang, W., Mao, H.-K, Hu, Q., Liu, Z., Schaller R. D., Zhang, D., and Xu, T. (2017). Pressure-induced bandgap optimization in lead-based perovskites with prolonged carrier lifetime and ambient retainability. *Adv. Funct. Mater.*, **27**, 1604208.

28. Frost, J. M., and Walsh, A. (2016). What is moving in hybrid halide perovskite solar cells?. *Acc. Chem. Res.*, **49**, 528–535.

29. Druzbicki, K., Pinna, R. S., Rudic, S., Jura, M., Gorini, G., and Fernandez-Alonso, F. (2016). Unexpected cation dynamics in the low-temperature phase of methylammonium lead iodide: the need for improved models. *J. Phys. Chem. Lett.*, **7**, 4701–4709.

30. Even, J., Pedesseau, L., Jancu, J.-M., and Katan, C. (2013). Importance of spin-orbit coupling in hybrid organic/inorganic perovskites for photovoltaic applications. *J. Phys. Chem. Lett.*, **4**, 2999–3005.

31. Mosconi, E., Amat, A., Nazeeruddin, M. K., Gratzel, M., and De Angelis, F. (2013). First-principles modeling of mixed halide organometal perovskites for photovoltaic applications. *J. Phys. Chem. C*, **117**, 13902–13913.

32. Kim, M., Im, J., Freeman, A. J., Ihm, J., and Jin, H. (2015). Switchable S = 1/2 and J = 1/2 rashba bands in ferroelectric halide perovskites. *PNAS.*, **111**, 6900–6904.

33. Zheng, F., Tan, L. Z., Liu, S., and Rappe, A. M. (2015). Rashba spin-orbit coupling enhanced carrier lifetime in $CH_3NH_3PbI_3$. *Nano Lett.*, **15**, 7794–7800.

34. Mosconi, E., Etienne, T., and De Angelis, F. (2017). Rashba band splitting in organohalide lead perovskites: bulk and surface effects. *J. Phys. Chem. Lett.*, **8**, 2247–2252.

35. Kepenekian, M., and Even, J. (2017). Rashba and dresselhaus couplings in halide perovskites: accomplishments and opportunities for spintronics and spin-orbitronics. *J. Phys. Chem. Lett.*, **8**, 3362–3370.

36. Kirchartz, T., and Rau, U. (2017). Decreasing radiative recombination coefficients via an indirect band gap in lead halide perovskites. *J. Phys. Chem. Lett.*, **8**, 1265–1271.

37. Zheng, C., and Rubel, O. (2017). Ionization energy as a stability criterion for halide perovskites. *J. Phys. Chem. C*, **121**, 11977–11984.

38. Miyata, A., Mitioglu, A., Plochocka, P., Portugall, O., Wang, J. T. W., Stranks, S. D., Snaith, H. J., and Nicholas, R. J. (2015). Direct measurement of the exciton binding energy and effective masses for charge carriers in organic-inorganic tri-halide perovskites. *Nat. Phys.*, **11**, 582–587.

39. Chen, Y., Peng, J., Su, D., Chen, X., and Liang, Z. (2015). Efficient and balanced charge transport revealed in planar perovskite solar cells. *ACS Appl. Mater. Interfaces*, **7**, 4471–4475.

40. Yin, W.-J., Yang, J.-H., Kang, J., Yan, Y., and Wei, S.-H. (2015). Halide perovskite materials for solar cells: a theoretical review. *J. Mater. Chem. A*, **3**, 4471–4475.

41. Yin, W.-J., Shi, T., and, Yan, Y. (2015). Unique properties of halide perovskites as possible origins of the superior solar cell performance. *Adv. Mater.*, **26**, 8926.

42. Uratani, H., and Yamashita, K. (2017). Charge carrier trapping at surface defects of perovskite solar cell absorbers: a first-principles study. *J. Phys. Chem. Lett.*, **8**, 742–746.

43. Sherkar, T. S., Momblona, C., Gil-Escrig, L., Avila, J., Sessolo, M., Bolink, H. J., and Koster, L. J. A. (2017). Recombination in perovskite solar cells: significance of grain boundaries, interface traps, and defect ions. *ACS Energy Lett.*, **2**, 1214–1222.

44. Ball, J. M., and Petrozza, A. (2016). Defects in perovskite-halides and their effects in solar cells. *Nature Energy*, **1**, 16149.

45. Sutter-Fella, C. M., Miller, D. W., Ngo, Q. P., Roe, E. T., Toma, F. M., Sharp, I. D., Lonergan, M. C., and Javey, A. (2017). Band tailing and deep states in $CH_3NH_3Pb(I_{1-x}Br_x)_3$ perovskites as revealed by sub-bandgap photocurrent. *ACS Energy Lett.*, **2**, 709–715.

46. Brenes, R., et al. (2017). Metal halide perovskite polycrystalline films exhibiting properties of single crystals. *Joule*, **1**, 155–167.

47. Yin, W.-J., Shi, T., and Yan, Y. (2014). Unusual defect physics in $CH_3NH_3PbI_3$ perovskite solar cell absorber. *Appl. Phys. Lett.*, **104**, 063903.

48. Yin, W.-J., Shi, T., and Yan, Y. (2015). Superior Photovoltaic Properties of Lead Halide Perovskites: Insights from First-Principles Theory. *J. Phys. Chem. C*, **119**, 5253–5264.

49. Zohar, A., Levine, I., Gupta, S., Davidson, O., Azulay, D., Millo, O., Balberg, I., Hodes, G., and Cahen, D. (2017). What is the mechanism of $MAPbI_3$ p-doping by I_2? Insights from optoelectronic properties. *ACS Energy Lett.*, **2**, 2408–2414.

50. Adinolfi, V., Yuan, M., Comin. R., Thibau, E. S., Shim D., Saidaminov, M. I., Kanjanaboos, P., Kopilovic, D., Hoogland, S., Lu, Z.-H., Bakr, O. M., and Sargent, E. H. (2016). The in-gap electronic state spectrum of methylammonium lead iodide sing-crystal perovskites. *Adv. Mater.*, **28**, 3406–3410.

51. Loper, P., Stuckelberger, M., Niesen, B., Werner, J., Filipic, M., Moon, S.-J., Yum, J.-H., Topic, M., De Wolf, S., and Ballif, C. (2015). Complex refractive index spectra of $CH_3NH_3PbI_3$ perovskite thin films determined by spectroscopic ellipsometry and spectrophotometry. *J. Phys. Chem. Lett.*, **6**, 66–71.

52. Almond, D. P., and Bowen, C. R. (2015). An explanation of the photoinduced giant dielectric constant of lead halide perovskote solar cells. *J. Phys. Chem. Lett.*, **6**, 1736–1740.

53. Miyata, K., Meggiolaro, D., Tuan Tinh, M., Joshi, P. P., Mosconi, E., Jouns, S. C., De Angelis, F., and Zhu, X.-Y. (2014). Large polarons in lead halide perovskites. *Sci. Adv.*, **3**, e1701217.

54. Hill, A. H., Smyser, K. E., Kennedy, C. L., Massaro, E. S., and Grumstrup, E. M. (2017). Screened charge carrier transport in methylammonium lead iodide perovskite thin films. *J. Phys. Chem. Lett.*, **8**, 948–953.

55. Kan, J., and Wang, L.-W. (2017). High defect tolerance in lead halide perovskite $CsPbBr_3$. *J. Phys. Chem. Lett.*, **8**, 489–493.

56. Li, W., Liu, J., Bai, F.-Q., Zhang, H.-X., and Prezhdo, O. V. (2017). Hole trapping by iodine interstitial defects decreases free carrier losses in perovskite solar cells: a time-domain *ab initio* study. *ACS Energy Lett.*, **2**, 1270–1278.

57. Brenner, T. M., Egger, D. A., Kronik, L., Hodes, G., and Cahen, D. (2016). hybrid organic-inorganic perovskites: low-cost semiconductors with intriguing charge-transport properties. *Nat. Rev.*, **1**, 15007.

58. Jankowska, J., Long, R., and Prezhdo, O. V. (2017). Quantum dynamics of photogenerated charge carriers in hybrid perovskites: dopants, grain boundaries, electric order, and other realistic aspects. *ACS Energy Lett.*, **2**, 1588–1597.

59. Stranks, S., Eperon, G. E., Grancini, G., Menelaou, C., Alcocer, M. J. P., Leijtens, T., Herz, L. M., Pertozza, A., and Snaith, H. J. (2013). Electron-hole diffusion lengths exceeding 1 micrometer in an organometla trihalide perovskite absorber. *Science*, **342**, 341–344.

60. Saba, M., Quochi, F., Mura, A., and Bongiovanni, G. (2016). Excited states properties of hybrid perovskites. *Acc. Chem. Res.*, **49**, 166–173.

61. Johnston, M. B., and Herz, L. M. (2016). Hybrid perovskites for photovoltaics: charge-carrier recombination, diffusion, and radiative efficiencies. *Acc. Chem. Res.*, **49**, 146–154.

62. D'Innocenzo, V., Grancini, G., Alcocer, M. J. P., Kandada, A. R. S., Stranks, S. D., Lee, M. M., Lanzani, G., Snaith, H. J., and Petrozza, A. (2014).

Excitons versus free charges in organo-lead tri-halide perovskites. *Nat. Commun.*, **5**, 3586.

63. Yu, Z.-G. (2017). Excitons in orthorhombic and tetragonal hybrid organic-inorganic perovskites. *J. Phys. Chem. C*, **121**, 3156–3160.

64. Grancini, G., Kandada, A. R. S., Frost, J. M., Barker, A. J., De Bastiani, M., Gandini, M., Marras, S., Lanzani, G., Walsh, A., and Petrozza A. (2015). Role of microstructure in the electron-hole interaction of hybrid lead halide perovskites. *Nat. Phys.*, **9**, 695–701.

65. Ham, S., Choi, Y. J., Lee, J.-W., Park, N.-G., and Kim, D. (2017). Impact of excess CH_3NH_3I on free carrier dynamics in high-performance nonstoichiometric perovskites. *J. Phys. Chem. C*, **121**, 3143–3148.

66. Hutter, E. M., Gelvez-Rueda, M. C., Osherov, A., Bulovic, V., Grozema, F. C., Stranks, S. D., and Savenije, T. J. (2016). Direct-indirect character of the bandgap in methylammonium lead iodide perovsite. *Nat. Mater.*, **16**, 115–121.

67. Manger, L. H., Rowley, M. B., Fu, Y., Foote, A. K., Rea, M. T., Wood, S. L., Jin, S., Wright, J. C., and Goldsmith, R. H. (2017). Global analysis of perovskite photophysics reveals importance of geminate pathways. *J. Phys. Chem. C*, **121**, 1062–1071.

68. Yang, Z., Surrente, A., Galkowski, K., Bruyant, N., Maude, D. K., Haghighirad, A. A., Snaith, H. J., Plochocka, P., and Nicholas, R. J. (2017). Unraveling the exciton binding energy and the dielectric constant in single-crystal methylammonium lead triiodide perovskite. *J. Phys. Chem. Lett.*, **8**, 1851–1855.

69. Jishi, R. A., Ta, O. B., and Sharif, A. A. (2014). Modeling of lead halide perovskites for photovoltaic applications. *J. Phys. Chem. C*, **118**, 28344–28349.

70. Herz, L. M. (2017). Charge-carrier mobilities in metal halide perovskites: fundamental mechanisms and limits. *ACS Energy Lett.*, **2**, 1539–1548.

71. Kang, B., and Biswas, K. (2017). Preferential $CH_3NH_3^+$ alignment and octahedral tilting affect charge localization in cubic phase $CH_3NH_3PbI_3$. *J. Phys. Chem. C*, **121**, 8319–8326.

72. Ma, J., and Wang, L.-W. (2017). The nature of electron mobility in hybrid perovskite $CH_3NH_3PbI_3$. *Nano Lett.*, **17**, 3646–3654.

73. Vrucinic, M., Matthiesen, C., Sadhanala, A., Divitini, G., Cacovich, S., Dutton, S. E., Ducati, C., Atature, M., Snaith, H. J., Friend, R. H., Sirringhaus, H., and Deschler, F. (2015). Local versus long-range diffusion effects of

photoexcited states on radiative recombination in organic-inorganic lead halide perovskites. *Adv. Sci.*, **2**, 1500136.

74. Zhu, H., Trinh, M. T., Wang, J., Fu, Y., Joshi, P. P., Miyata, K., Jin, S., and Zhu, X.-Y. (2016). Organic cations might not be essential to the remarkable properties of band edge carriers in lead halide perovskies. *Adv. Mater.*, **29**, 1603072.

75. Brenner, T. M., Egger, D. A., Rappe, A. M., Kronik, L., Hodes, G., and Cahen, D. (2015). Are mobilities in hybrid organic-inorganic halide perovskites actually "high"?. *J. Phys. Chem. Lett.*, **6**, 4754–4757.

76. Bonn, M., Miyata, K., Hendry, E., and Zhu, X.-Y. (2017). Role of dielectric drag in polaron mobility in lead halide perovskites. *ACS Energy Lett.*, **2**, 2555–2562.

77. Miyata, K., Atallah, T., and Zhu, X.-Y. (2017). Lead halide perovskites: crystal-liquid duality, phonon glass electron crystals, and large polaron formation. *Sci. Adv.*, **3**, e1701469.

78. Karakus, M., Jensen, S. A., D'Angelo, F., Turchinovich, D., Bonn, M., and Canovas, E. (2015). Phonon-electron scattering limits free charge mobility in methylammonium lead iodide perovskites. *J. Phys. Chem. Lett.*, **6**, 4991–4996.

79. Zhu, X.-Y., and Podzorov, V. (2015). Charge carriers in hybrid organic inorganic lead halide perovskites might be protected as large polarons. *J. Phys. Chem. Lett.*, **6**, 4758–4761.

80. Ivanovska, T., Dionigi, C., Mosconi, E., De Angelis, F., Liscio, F., Morandi, V., and Ruani, G. (2017). Long-lived photoinduced polarons in organohalide perovskites. *J. Phys. Chem. Lett.*, **8**, 3081–3086.

81. Mante, P.-A., Stoumpos, C. C., Kanatzidis, M. G., and Yartsev, A. (2017). Electron-acoustic phonon coupling in single crystal $CH_3NH_3PbI_3$ perovskites revealed by coherent acoustic phonons. *Nat. Commun.*, **8**, 14398.

Chapter 3

Optical Excited-State Properties of Halide Perovskites

Valerio Sarritzu, Nicola Sestu, Daniela Marongiu, Xueqing Chang, Francesco Quochi, Michele Saba, Andrea Mura, and Giovanni Bongiovanni
Dipartimento di Fisica, Università degli Studi di Cagliari, SP Monserrato-Sestu km 0,700, 09042 Monserrato CA, Italy
valerio.sarritzu@dsf.unica.it

In keeping with the nature of hybrid materials, halide perovskites have been found to possess an attractive blend of optoelectronic properties. They exhibit strong, sharp optical absorption, high charge carrier mobilities, long diffusion lengths, and low trap density typical of inorganic, epitaxial semiconductors [1–3] and yet they can be solution-processed into high-quality thin films at a low cost, like organic semiconductors [4–6]. It comes as no surprise then that the scientific community has placed a lot of expectations on this new class of materials as a less expensive alternative to conventional active media in a variety of applications, such as solar cells and photonic sources [7–10]. The importance of understanding the excited-state properties of such a promising material cannot

Multifunctional Organic–Inorganic Halide Perovskite: Applications in Solar Cells, Light-Emitting Diodes, and Resistive Memory
Edited by Nam-Gyu Park and Hiroshi Segawa

ISBN 978-981-4800-52-5 (Hardcover), 978-1-003-27593-0 (eBook)
www.jennystanford.com

be overstated, because its behavior in optoelectronic devices is determined by the nature of band-edge transitions. In this chapter we offer a summary of the most relevant excited-state properties of halide perovskites.

3.1 Excitons versus Free Carriers

Assessing whether photon absorption yields a population of excitons or unbound electrons and holes is of primary importance from both fundamental and applied standpoints. The formation of unbound electron-hole pairs upon photoexcitation results in a conductive plasma of free charges. Excitons, on the other hand, form an insulating gas of neutral particles that would require a heterojunction to be split so the charge can be harvested. Either way, advances in the engineering of optoelectronic devices based on halide perovskites rely on understanding of the role of excitons, whether it is marginal, as in inorganic semiconductors, or crucial, like in organics. Here, we review the most significant photophysical evidence, concluding that halide perovskites behave as free-charge semiconductors.

3.1.1 Exciton Binding Energy

The exciton binding energy is a key parameter in the optoelectronics of direct-gap semiconductors. It determines the stability of excited electron-hole pairs against thermal dissociation and deeply affects the design and workings of a device. Consensus is still lacking on how large it actually is in halide perovskites, even for the most widely studied varieties, such as methylammonium lead iodide ($MAPbI_3$) and methylammonium lead bromide ($MAPbBr_3$). Widely spread values for the binding energy have been reported, for example, ranging from 2 to 50 meV in $MAPbI_3$ [11–20].

The reason for this wide range of values has been found [21] in a shortcoming of the customary method for estimating the exciton binding energy, which can be summarized as follows. The first step is to measure the optical absorption spectrum near the bandgap. Its shape can be described by the Elliott formula [22], convoluted with a bell-shaped function $g(E)$ in order to account for the line width Γ [23, 24]:

$$\alpha(\hbar\omega) \propto \mu_{\text{cv}}^2 \sqrt{E_{\text{B}}} \left(\sum_n \alpha_{\text{nx}} + \alpha_{\text{c}} \right)$$

$$\propto \mu_{\text{cv}}^2 \sqrt{E_{\text{B}}} \left[\sum_n \frac{2E_{\text{B}}}{n^3} g\left(\frac{\hbar\omega - E_{nx}}{\Gamma} \right) \right. \tag{3.1}$$

$$\left. + \int_{E_{\text{g}}}^{\infty} g\left(\frac{\hbar\omega - E}{\Gamma} \right) \frac{1 + b(E - E_{\text{g}})}{1 - e^{-2\pi\sqrt{\frac{E_{\text{B}}}{E - E_{\text{g}}}}}} \text{d}E \right]$$

Here, $\alpha_{n\text{x}}$ is the contribution from the n-exciton state with energy $E_{n\text{x}} = E_{\text{g}} - (E_{\text{B}}/n^2)$ and binding energy (E_{B}/n^2), α_{c} is the contribution from continuum states, while $\mu_{\text{cv}}{}^2$ is the transition dipole moment. The term $b(E - E_{\text{g}}) = 10(m^2E/\hbar^4)c_{\text{np}} + 126\ ((m^2E/\hbar^4)c_{\text{np}})^2$ accounts for nonparabolic conduction and valence bands, assuming a dispersion relation $E(k) = (\hbar^2k^2/2m)c_{\text{np}}k^4$. An estimate for E_{B} may be then obtained from a least-squares fit of Eq. 3.1 to the absorption spectrum. However, the accuracy and sensitivity of the method depend on the excitonic peak being spectrally resolved, which is the case at cryogenic temperatures. At room temperature, the line width and exciton binding energy are similar in magnitude and the procedure fails to disentangle the exciton and continuum contributions.

A different approach [24] overcame this issue by relying on quantities that can be extracted from the raw experimental data, without the need for a least-squares fit. This paragraph summarizes the method and its results. Figure 3.2 shows a series of absorption spectra for a $MAPbBr_3$ film, acquired at different temperatures. As a first step, they have been normalized at an energy E_{n} within the continuum in order to cancel out the dependence on μ_{cv}, together with the linear dependence on E_{B} of the prefactor in the absorption coefficient. This preserves the information on the relative contributions of the exciton and continuum transitions needed to monitor E_{B}. The next step is the evaluation of the quantity:

$$I = \frac{1}{\alpha(E)} \int_0^{E_{\text{n}}} \alpha(E)\text{d}E \tag{3.2}$$

which is the area under the normalized absorption curve up to the normalization energy. Thanks to the normalization, I does not depend on Γ or on the overall transition dipole moment μ_{cv}. The normalization energy E_n and the nonparabolicity coefficient c_{np} affect the value of I as parameters, so I is a function of E_B only. In other words, I does not depend on film thickness, energy bandgap, or line width.

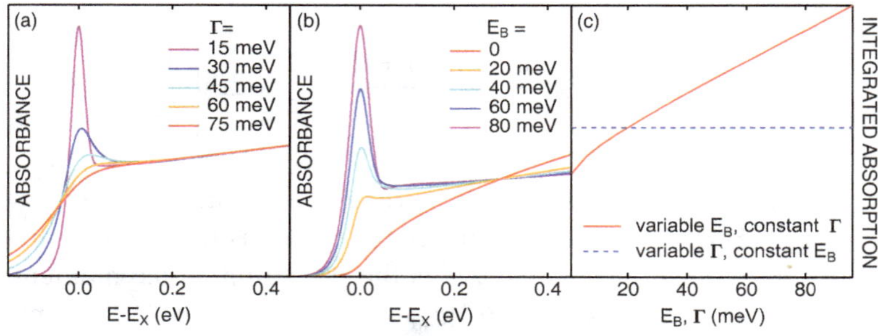

Figure 3.1 UV-Vis absorption spectra computed according to the Elliott formula with (a) the line width Γ and (b) the exciton binding energy E_B as variables. Numerical values for the parameters in the computation were chosen to represent those typical of $MAPbBr_3$, that is E_B = 50 meV in panel (a), Γ = 20 meV in panel (b), and E_g = 2.3 eV in all three of the panels. Spectra are plotted versus the energy difference with respect to the n = 1 exciton transition and normalized at their value for $E_n = E_{1x}$ + 0.3 eV. Panel (c) shows that the integral of the normalized absorption spectra $I = (1/\alpha(E_n)) \int E_n\alpha(E)dE$ grows as a function of the exciton binding energy E_B (red line), while it is constant when only the line width Γ varies (blue line).

Figure 3.1 illustrates the role of the line width and the exciton binding energy in shaping the absorption coefficient. Spectra have been computed according to Eq. 3.1. In Fig. 3.1a, the line width Γ varies but the exciton binding energy E_B is kept constant. The exciton peak broadens out, but both the area under the exciton peak and the slope of the continuum contribution to the absorbance do not change, owing to the fact that broadening does not modify the oscillator strength of the optical transitions. When E_B varies (Fig. 3.1b), the ratio between the exciton peak and the continuum, as well as the slope of the continuum contribution, varies between two extremes: if $E_B \ll \Gamma$, the absorption shows a plateau on the high-energy side of the exciton peak, because the socalled Sommerfeld enhancement factor decreases as $1/(\hbar\omega)^{1/2}$ and compensates for the $(\hbar\omega)^{1/2}$ growth of the density of states (the upward slope is mostly due to band nonparabolicity); in the opposite case, of vanishingly small exciton binding energy, absorbance grows as the square root of energy ($\alpha(\hbar\omega) \propto (\hbar\omega)^{1/2}$), as predicted by the calculation of the joint density of states for non-interacting electron and holes.

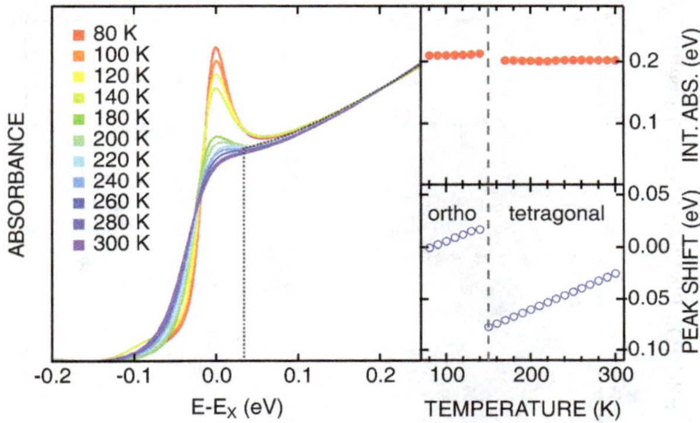

Figure 3.2 Analysis of the experimental absorption spectra for $MAPbI_3$ films. (a) UV-Vis absorption spectra for different temperatures, normalized at the energy $E_n = E_{1x} + 0.2$ eV. The gray dotted line is the continuum contribution α_c calculated according to Elliott formula with $(m^2/\hbar^4)c_{np} = 0.1$ eV^{-1} and $\Gamma = 0$. Spectra are shown with 20 K temperature interval for clarity purposes, although measurements were taken at 10 K intervals. (b) Normalized absorbance integrated up to energy E_n (defined as $I = (1/\alpha(E_n)) \int E_n \alpha(E)dE$ in the text) as a function of temperature. (c) Shift of the excitonic peak in the absorption spectra as a function of temperature; when the exciton peak was not well resolved in absorption, the energy of the photoluminescence (PL) peak was taken instead, taking advantage of the negligible Stokes shift. The vertical dashed line marks the temperature of the phase transition.

Figure 3.1c sums it all up, showing that the integrated normalized absorption I does not depend on temperature when the exciton binding energy is constant with temperature. It does increase, though, when E_B increases. Therefore, it can be used to monitor the exciton binding energy: if on increasing the temperature I stays constant, it means that the temperature just broadens the transitions, without changing E_B; if instead E_B changes with temperature, then corresponding changes in I have to be measured. The main experimental findings are shown in Fig. 3.3 and can be summarized as follows. Apart from the discrete jump at the phase transition in $MAPbI_3$, there is no evidence of temperature dependence of the exciton binding energy. In $MAPbI_3$, $E_B = (34 \pm 3)$

meV in the orthorhombic phase (up to 140 K) and E_B = (29 ± 3) meV from 170 K to 300 K. The exciton binding energy is E_B = (60 ± 3) meV in $MAPbBr_3$, with no dependence on temperature between 80 K and 300 K.

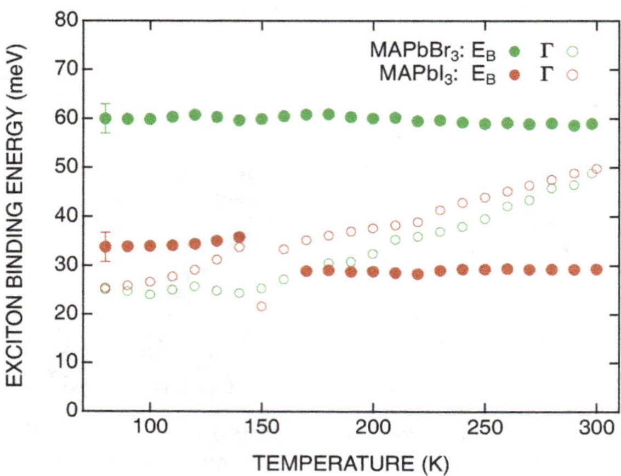

Figure 3.3 Exciton binding energy versus temperature in $MAPbI_3$ (filled red circles) and $MAPbBr_3$ (filled green circles) films. The error bars are given on the first point for each material. Empty circles are the line width as extracted from the absorption spectra with the formula $\Gamma = \int_0^{E_{1x}} \alpha(E)dE)/(\alpha(E_{1x})$.

3.1.2 Saha Equilibrium

The exciton binding energy allows making a quantitative prediction about the quasi-equilibrium between bound and unbound electron-hole pairs in a semiconductor. The thermodynamic equilibrium ratio between exciton and free carrier population densities (n_x and $n_{e,h}$) at any given temperature and overall excited density is expressed by the mass action law known as the Saha equation [25, 26]:

$$\frac{n_{e,h}^2}{n_x} = \left(\frac{\mu_x kT}{2\pi\hbar^2}\right)^{3/2} e^{-\frac{E_B}{kT}} = n_{eq} \tag{3.3}$$

where μ_x = 0.15 m_e is the effective reduced mass of the exciton, m_e being the electron mass, and T is the temperature of the optical excitations [27]. The resulting prediction can be validated by the means of time-resolved PL measurements, and this will be discussed in Section 3.1.3. Plugging the values for E_B from Section 3.1.1 into Eq. 3.3, the result is that at room temperature free carriers vastly outnumber excitons in both $MAPbI_3$ and $MAPbBr_3$ up to carrier densities around 10^{17} cm^{-3}.

The question is still open as to what happens at cryogenic temperatures. The formation and ionization of the excitons occur at a rate that depends on both carrier and lattice temperature through interaction with optical and acoustic phonons. In fact, the system does not reach thermodynamical equilibrium at all. The formation rate in halide perovskites is yet to be assessed. The time needed for excitons to form might lead to never reaching chemical equilibrium between the two photoexcited species. This was, for example, the case in InGaAs quantum wells. Time-resolved luminescence experiments carried out on those structures showed that, except for high excitation densities, the process of exciton formation is slow compared to the radiative decay of excitons [28].

3.1.3 The Spectroscopic Signature of Free Carriers

A comprehensive picture of the nature of photoexcited species and the dynamics of excited-state relaxation can be inferred from time-resolved PL measurements. Subpicosecond laser pulses are able to inject a population of excited carriers instantaneously with respect to the timescale of the decay processes that such carriers undergo. The subsequent temporal evolution of the PL can then be tracked by various means. The evidence collected by transient PL, differential transmission experiments, and transient THz spectroscopy points to the fact that free charges are the dominant species even at excitation densities higher than 10^{15} cm^{-3} [29, 30–32].

A lot can be understood by looking at the initial behavior of the PL profile. Figure 3.4a shows the transient PL signal for a $MAPbI_3$ thin film under various excitation intensities. The PL signal rises instantaneously, meaning that the exciton population—if there is any at all—is formed within the temporal resolution of the experimental setup, which is 60 ps [29]. Right after excitation, photoexcited electrons and holes quickly thermalize to the edges of their respective bands.

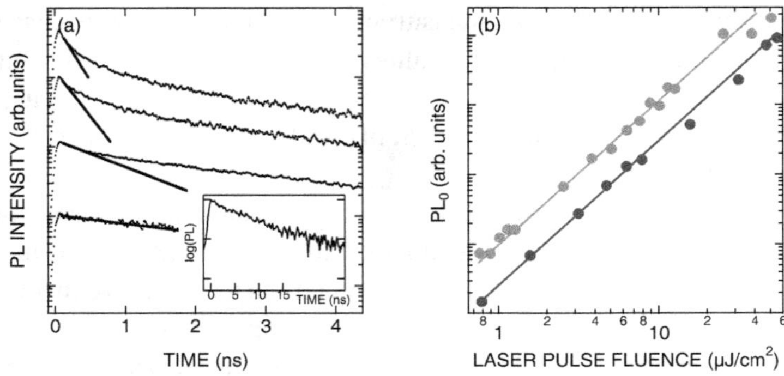

Figure 3.4 Time-resolved photoluminescence from a $MAPbI_3$ thin film. (a) Transient photoluminescence signal at various injected electron-hole pair densities at the film surface. Straight lines in the semilogarithmic plot represent an exponential fit to the initial decay of the photoluminescence signal. Photoluminescence was excited by 150 fs long laser pulses with a repetition rate of 1 kHz and 3.18 eV photon energy. Injected carrier densities, from top to bottom: $1.2 \cdot 10^{19}$, $3.9 \cdot 10^{18}$, $1.2 \cdot 10^{18}$, and $1.9 \cdot 10^{17}$ cm^{-3}. (b) Photoluminescence emission intensity estimated at time $t = 0$ after excitation (PL_0) as a function of laser pulse fluence. The quadratic dependence is shown by the line as a guide for the eye.

The dependence of the recombination rate on the injection carrier density has a noticeable effect on the transient PL signal. The exponential fits in Fig. 3.4a show that the initial decay time becomes shorter as the excitation density increases, indicating the activation of density-dependent recombination mechanisms. This can be understood by looking at the rate equation for the time-dependent carrier density n:

$$\frac{dn}{dt} = -k_1 n - k_2 n^2 - k_3 n^3 \tag{3.4}$$

where k_1, k_2, and k_3 are the rate constants for monomolecular, bimolecular, and Auger recombination, respectively. At low excitation densities, the decay is nearly exponential, indicating monomolecular recombination induced by charge trapping. Raising the injection level yields faster decays, as the importance of the bimolecular radiative recombination of electron-hole pairs increases. At very high densities, it becomes even faster due to the growing relevance of Auger processes, scaling as n^3. Later on, decays are exponential regardless of the injected population. Indeed, the time-dependent carrier density eventually falls back to levels where monomolecular processes prevail.

Another quantity of interest is the value of the PL signal right after excitation, here labeled PL_0, just before electronic states are depopulated by slow recombination processes. PL_0 is proportional to the rate of spontaneous photon emission per unit of volume, $R(n)$, which in turn depends on the injected carrier density n_0. Figure 3.4 shows PL_0 scales as the square of the density of optically injected carriers. Such scaling is the signature of bimolecular emission by a gas of unbound electron-hole pairs, where the radiative recombination rate is proportional to the probability of an electron meeting a hole and therefore to the product of electron and hole densities: $R(n) \propto n_e n_h = n^2{}_0$. Remarkably, the very same behavior dependence is also observed in the wide-bandgap perovskite $MAPbBr_3$, validating the prediction of free carriers from the Saha equation for its $E_B \approx 60$ eV $>> kT$ [29].

3.1.4 Technological Implications

As mentioned earlier, the nature of excited-state species determines how a semiconductor can suit different applications. Free carrier semiconductors favor charge transport and harvesting in solar cells. In addition to that, thanks to a bandgap that can be tuned over the entire solar spectrum, halide perovskites are very promising materials for single- and multijunction solar cells. On the other hand, their use as color-tunable materials in light-emitting devices might be hampered by the fact that radiative recombination between free carriers is a bimolecular process, with very low efficiency at the injected densities required for light-emitting diode (LED) operation. Such limitation could be addressed by designing nanostructured perovskite materials with carrier confinement (see Section 3.3) and enhanced exciton binding energies. On the other hand, the very same process becomes very efficient with a quantum yield close to unity at population densities close to inversion [29]. Bimolecular recombination could be embraced, employing the material as a gain medium in lasers, as several works have proposed [30, 33–38].

3.2 Electron–Hole Recombination in Perovskites and Perovskite-Based Solar Cells

Halide perovskite solar cells rival the best inorganic solar cells in power conversion efficiency. Devices based on $MAPbI_3$ saw a surge to over 20% within the span of just a few years [39]. High-quality perovskite thin films can be formed at a low temperature using inexpensive techniques, such as vapor deposition in vacuum or spin coating, making them a viable option as third-generation light-harvesting materials. This triggered a considerable effort in the study of the photophysics of this class of materials. In particular, a detailed understanding has been long sought of the recombination processes that charge carriers undergo in perovskite and perovskite-based solar cells.

3.2.1 *I–V* Characteristics

Customarily, the electrical characteristics of solar cells are studied to tell apart the various processes of energy loss. This can be done in a number of ways, one of which typically involves measuring the *I–V* (current-voltage) characteristics of the cell in the dark. The basic model for the device would then be that of a diode. However, the ideal diode equation would only apply if all electron–hole pairs decayed radiatively. If this were the case, power conversion efficiency would hit its theoretical upper bound, also referred to as the Shockley–Queisser limit [40]. A more general expression accounting for different, nonradiative recombination channels can be obtained by introducing the ideality factor, m, and reads as follows:

$$I = I_0\left[\exp\left(\frac{qV}{mkT}\right) - 1\right] \tag{3.5}$$

where I is the current through the device, V is the voltage across it, I_0 is the dark saturation current, e is the elementary charge, and T is the temperature. If the contribution from the dark saturation current can be neglected, Eq. 3.5 can be conveniently rewritten as:

$$\ln(I) = \ln(I_0) + \left(\frac{qV}{mkT}\right)V \tag{3.6}$$

so that the ideality factor can be derived from the slope of the *I–V* characteristics. Information on which recombination process dominates can be inferred from the value of *m*.

The method has yet to provide conclusive insight. So far, the values reported for halide perovskites mainly lie in the range 1.7–2 [41–44] but extend as far up as 5 [45]. Given the logarithmic dependence of the cell voltage on the recombination rate, the determination of the ideality factor is not very reliable, especially for halide perovskite solar cells, where hysteresis and degradation effects may lead to distortion of the *I–V* characteristics [46–50]. As a consequence, it is difficult to identify from the electrical characterization what recombination processes limit the photoconversion efficiency and, therefore, to elaborate an informed strategy to improve the devices. The following section details how to determine the ideality factor from an all-optical experiment and identify the prevailing recombination channels. Instead of the *I–V* curves, the method focuses on the free energy of the electron–hole pairs (μ) as a function of the intensity of the exciting light (I_{ex}), namely the μ–I_{ex} characteristics. The analysis shows that Shockley–Read–Hall (SRH) nonradiative recombination dominates in single perovskite layers, while interface recombination is the main nonradiative process in single and double heterojunctions [51].

3.2.2 All-Optical Determination of the Ideality Factor

In a solar cell, the charge available for harvesting during operation is determined by the steady-state balance between generation and recombination processes in the absorber. According to the theory of nonequilibrium semiconductors, photoexcitation results in a splitting of the quasi-Fermi levels of electrons in the conduction and valence bands, respectively. As shown in Fig. 3.5, this splitting is the free energy of photoexcited carriers μ. In the simplest architecture of a heterojunction solar cell, the absorber is sandwiched between two charge-selective semiconductor layers, one of which (electron transport layer [ETL]) allows photoexcited electrons to flow but blocks the transmission of holes, while the other (hole transport layer [HTL]) only lets photoexcited holes reach the opposite electrode (Fig. 3.5). If the quasi-Fermi levels do not vary from the transport materials up to the external contacts, the circuit voltage V of the solar cell is given by μ/e in the intrinsic layer, where e is the elementary charge.

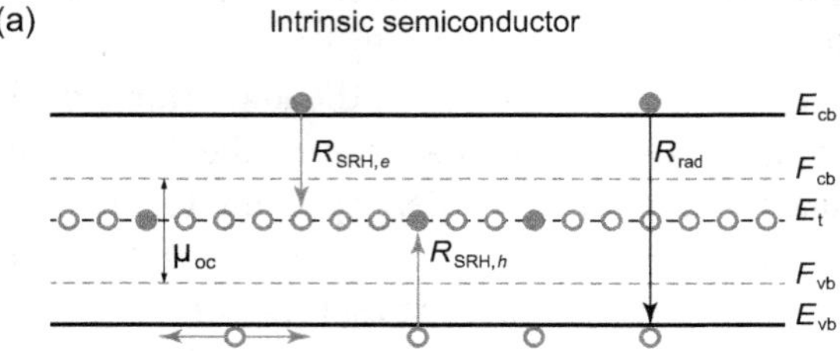

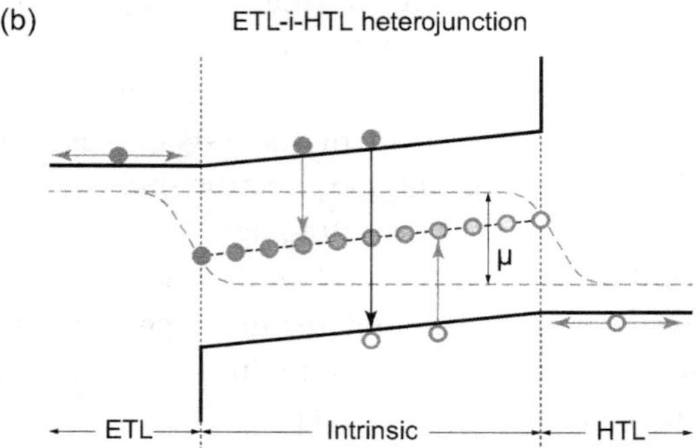

Figure 3.5 Electron and hole energetics and recombinations in perovskites and perovskite-based solar cells. (a) Stand-alone intrinsic layer. The free energy μ_{oc} of photogenerated electron–hole pairs is equal to the energy splitting of the quasi-Fermi levels of electrons in the conduction (F_{cb}) and valence (F_{vb}) bands. No electric voltage is present between the two sides due to the absence of electron- and hole-selective contacts. An empty trap level E_t in the midgap is assumed. $R_{SRH,e(h)}$ is the Shockley–Read–Hall recombination rate of electrons (holes) per unit of volume. Correspondingly, R_{rad} refers to radiative recombinations. (b) HTL-i-ETL double heterojunction. The difference between F_{cb} in the ETL and F_{vb} in the HTL is given by μ = eV, where V is the circuit voltage of the solar cell in the absence of electrical losses. Equilibrium conditions of the solar cell in the dark impose that the concentration of trapped electrons varies across the intrinsic layer: traps in the center of the i-semiconductor are half-filled, so the recombination rate $R_{SRH,e(h)}$ of excess carriers is quite large. Conversely, traps close to the ETL (HTL) are filled (empty), so no trapping of electrons (holes) by midgap states is possible, leading to negligible values for $R_{SRH,e(h)}$.

μ can be expressed in terms of measurable optical quantities. According to Kirchhoff's law of radiation, which represents the detailed balance between emission and absorption, generalized by Wu¨rfel to account for nonequilibrium electron and hole populations:

$$J_{\mathrm{PL}} = \int \alpha(\omega) \frac{\Omega}{4\pi^3 \hbar^3 c^2} \frac{\hbar^3 \omega^2}{e^{\frac{\hbar\omega - \mu}{kT}} - 1} \mathrm{d}(\hbar\omega) \cong J_{0,\mathrm{rad}} e^{\frac{\mu}{kT}}, \tag{3.7}$$

where J_{PL} is the emitted photon current density, which is proportional to the external PL intensity; $\alpha(\omega)$ is the absorptivity, which depends both on the absorption coefficient and the thickness; ω is the angular frequency of the radiation; Ω is the effective external emission angle; c is the speed of light; k is the Boltzmann constant; and T is the temperature. The right-hand side of Eq. 3.7 holds when the Bose function can be approximated by the Boltzmann distribution, which is the case at excitation levels typical of solar illumination. Equation 3.7 provides the analytical relation between J_{PL}, μ, and the photon current density $J_{0,\mathrm{rad}}$ emitted by the semiconductor in thermal equilibrium with the environment ($\mu = 0$).

Further manipulation leads to a more convenient way to relate μ to measurable quantities. J_{PL} can be expressed in terms of the external PL quantum yield ($\mathrm{EQY_{PL}}$), defined as the ratio between J_{PL} and the absorbed excitation photon flux, that is $\mathrm{EQY_{PL}} = J_{\mathrm{PL}}/J_{\mathrm{ex}}$. Equation 3.7 can be reformulated to explicitly link μ to $\mathrm{EQY_{PL}}$:

$$\begin{aligned} \mu &= kT \ln\left(\frac{J_{\mathrm{PL}}}{J_{0,\mathrm{rad}}}\right) \\ &= kT\left[\ln\left(\frac{J_{\mathrm{ex}}}{J_{0,\mathrm{rad}}}\right) + \ln\left(\mathrm{EQY_{PL}}\right)\right] \\ &= \mu_{\mathrm{oc,rad}} + kT \ln\left(\mathrm{EQY_{PL}}\right) \end{aligned} \tag{3.8}$$

in which $\mu_{\mathrm{oc,rad}} = \ln(J_{\mathrm{ex}}/J_{0,\mathrm{rad}})$ represents the upper limit for the free energy when only radiative decays occur ($\mathrm{EQY_{PL}} = 1$) and in the open circuit condition. If all incident photons with energy higher than the optical gap are absorbed, $\mu_{\mathrm{oc,rad}}$ represents the Shockley–Queisser limit to the open circuit voltage V_{oc}. It is interesting to note that nonradiative decays affect the free energy only through $\mathrm{EQY_{PL}}$ and that a factor of 10 drop in $\mathrm{EQY_{PL}}$ yields a loss of free energy of $kT \ln (10^{-1}) = 60$ meV at 300 K.

While Eq. 3.8 allows measuring μ, it does not provide any information on the mechanisms of electron–hole recombination, which is the ultimate goal. The missing link is how the decay processes influence the μ-I_{ex} characteristics, and it can be found as follows. Let us consider the simplest case, in which decays of electrons and holes are driven by elementary processes such as monomolecular, bimolecular, or higher-order multiparticle interactions. Let α be the number of carriers involved in the recombination process: α = 1, 2, and 3 for monomolecular, bimolecular, and trimolecular recombinations, respectively. The corresponding recombination rates for electrons and holes are proportional to the αth power of the population densities through the appropriate coefficient k_α. If thermal excitation and recombination are negligible, the following relation holds under steady state:

$$k_\alpha n_{e(h)}^{\alpha} \simeq \frac{J_{ex} - J_{e(h)}}{d}, \tag{3.9}$$

where J_{ex}/d is the mean carrier generation rate per unit of volume, $J_{e(h)}/d$ is the electron (hole) extraction rate per unit of volume, and μ depends on the free electron (n_e) and hole (n_h) concentrations according to the mass action law, generalized to account for nonequilibrium carriers:

$$\mu = kT \ln\left(\frac{n_e n_h}{n_i^2}\right) \tag{3.10}$$

where n_i is the free carrier concentration of the intrinsic semiconductor in the dark. Defining

$$m = \frac{1}{\alpha_e} + \frac{1}{\alpha_h}, \tag{3.11}$$

Equation 3.10 can be rewritten as:

$$\begin{aligned}\mu &\cong \left(\frac{1}{\alpha_e} + \frac{1}{\alpha_h}\right) kT \ln\left(\frac{J_{ex} - J_{e(h)}}{J_0}\right) \\ &= mkT \ln\left(\frac{J_{ex} - J_{e(h)}}{J_0}\right)\end{aligned} \tag{3.12}$$

in which J_0 is a constant. For optoelectronic measurements, when a current is allowed to flow ($J_{e(h)} \neq 0$), Eq. 3.12 is often written by replacing μ = eV in the form $J_{e(h)} \cong J_{sc} - J_0$ exp(eV/mkT), where m

is the ideality factor of the μ-J_{ex} curves, with J_0 being the reverse bias saturation current (note again that charge currents are usually considered instead of particle currents, as we are doing here).

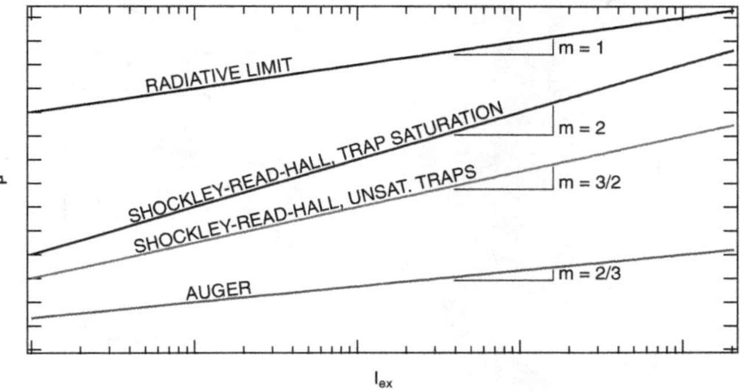

Figure 3.6 Sketch of the μ-J_{ex} characteristics. The ideality factor is given by the slope of the data in a log-log plot.

When $J_{e(h)} = 0(\mu = \mu_{oc})$, Eq. 3.12 provides an all-optical route to determine the ideality factor m and consequently to identify the recombination mechanisms. Elementary electron–hole annihilation processes (m = 1, 2, 3, . . .) are associated with rational values of m. Band-to-band electron–hole recombinations ($n_e \sim n_h$, $\alpha_e = \alpha_h = 2$) yield $m = 1$; nonradiative monomolecular decays of minority carriers in doped semiconductors ($\alpha_{e(h)} = 1, 1/\alpha_{h(e)}$) also lead to $m = 1$. Auger recombinations are trimolecular processes, so $m = 2/3$ ($n_e \sim n_h$, $\alpha_e = \alpha_h = 3$). SRH nonradiative decays in the space-charge region of a p-n junction or in the intrinsic layer of a p-i-n (HTL-i-ETL) device, for which $\alpha_e = \alpha_h \cong 1$, is instead characterized by $m = 2$. The ideality factor is just given by the slope of the μ-I_{ex} data (Fig. 3.6).

3.2.3 Shockley–Read–Hall Recombination in Perovskite Films

The upper limit for cell voltage is given by the recombination processes in the absorber. This is because the bulk carrier recombination processes determine the maximum attainable free energy μ_{oc}. Figure 3.7 reports μ_{oc} as a function of $\ln(I_{ex})$, where I_{ex} is the excitation light intensity delivered by a green continuous wave (CW) laser (λ = 532 nm). The free energy μ_{oc} was estimated through

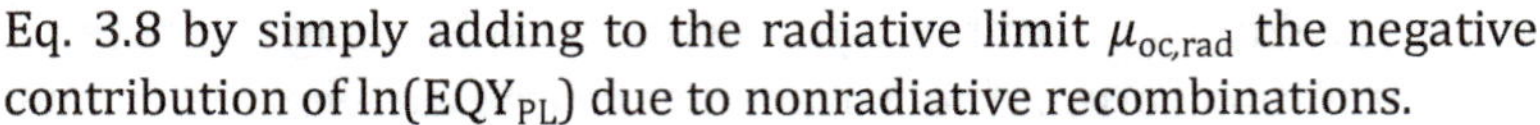

Eq. 3.8 by simply adding to the radiative limit $\mu_{oc,rad}$ the negative contribution of $\ln(EQY_{PL})$ due to nonradiative recombinations.

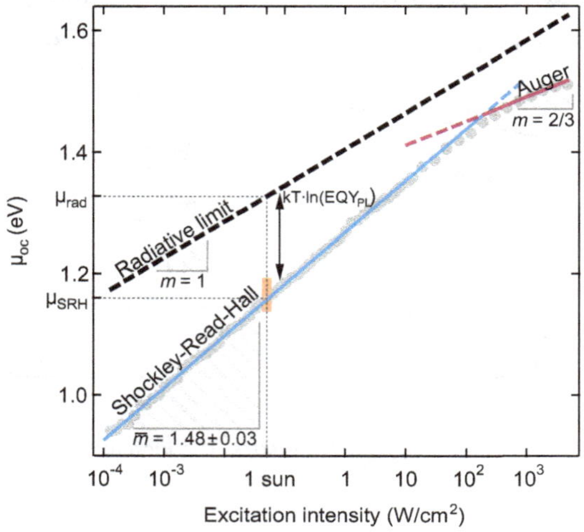

Figure 3.7 μ-J_{ex} characteristics of intrinsic $MAPbI_3$ films. Gray markers represent the measured free energy μ_{oc} as a function of the excitation intensity J_{ex} delivered by a CW laser at 532 nm. The orange box encloses the dispersion of μ_{oc} for samples fabricated with a variety of different techniques. For 1 mW/cm^2 < I_{ex} < 100 W/cm^2, μ_{oc} depends linearly on a slope m of approximately 3/2 regardless of the fabrication method, yielding an average ideality factor $\overline{m}$ = 1.48 ± 0.03 (within one standard deviation). Shockley–Read–Hall recombinations are expected to lead to a rational ideality factor m = 3/2 at low illumination levels. The experimental slope decreases to approximately 2/3 for I_{ex} > 100 W/cm^2, as foreseen for Auger recombination. At I_{ex} = 50 mW/cm^2, the rate of photons absorbed by the film matches that one obtained at an illumination level of 1 sun (AM 1.5 G). At this excitation intensity, μ_{rad} = 1.33 eV while $\overline{\mu}_{SRH}$ = 1.16 eV, with a free energy loss due to SRH recombinations $\Delta\mu_{SRH} = \mu_{rad} - \overline{\mu}_{SRH} = kT\ln(EQY_{PL})$ = 0.17 eV.

μ_{oc} was determined from the EQY_{PL} of $MAPbI_3$ as a function of I_{ex}. According to Eq. 3.8, the value of $\mu_{oc,rad}$ is needed to calculate the free energy. Both optical and electrical measurements are in agreement on a value that is safe to assume: $\mu_{oc,rad}(J_{sun})$ = 1.33 ± 0.02 eV. We analyzed single perovskite layers grown by several deposition methods, namely single-step and double-step techniques.

Experimental data show that μ is proportional to $\ln(I_{ex})$, with the slope coefficient giving the ideality factor. For 1 mW/cm^2 < I_{ex} < 100 W/cm^2, the slope was close to 3/2, independently of the method used to fabricate the $MAPbI_3$ films, with an average ideality factor $\overline{m} = 1.48 \pm 0.03$. This is an indication that electrons and holes follow different power laws and, more specifically, one carrier decays obeying a first-order process ($\alpha = 1$) while the other follows a second-order process ($\alpha = 2$). This asymmetry between electrons and holes is consistent with the presence of traps preferentially capturing a single carrier type. Figure 3.5 shows the recombination processes expected in a stand-alone, undoped HP layer, assuming empty traps; fully occupied trap states would lead to similar conclusions, just swapping the roles of electrons and holes. The capture rate of electrons by intragap levels can be described as a bimolecular process, $R_{SRH,e} \propto n_{h,t}n_e$, where $n_{h,t}$ is the density of trapped holes. As for holes, since all traps are nearly empty, $n_{h,t}$ is almost equal to the density N_t of recombination centers, so recombination is monomolecular ($\alpha_e = 1$). Similarly to the case of electrons, the hole capture rate is $R_{SRH,h} \propto n_h n_{e,t}$, but in this case, the densities of both trapped electrons $n_{e,t}$ and free holes n_h increase with the excitation intensity and, additionally, $n_{e,t} \approx n_h$. Hole recombination is therefore an effective bimolecular process with $\alpha_h = 2$; far from trap saturation, we obtain $m = 1/\alpha_e + 1/\alpha_h = 3/2$.

We thus conclude that nonradiative recombinations in halide perovskites can be described in the framework of the SRH model. In our samples, SRH recombination limits the available free energy to $\mu_{oc} = 1.16$ eV, with a loss at 1 sun excitation $\overline{\Delta\mu_{oc,SRH}} = \mu_{oc,rad} - \overline{\mu}_{oc} = kT \ln(\overline{EQY_{PL}}) = 0.177$ eV; lower losses could be achieved by reducing the trap density. For $I_{ex} > 100$ W/cm^2, m decreased to $\approx 2/3$, as expected for trimolecular annihilations via Auger decay channels. Similar behavior is also observed in the I–V characteristics of a Si solar cell, where m decreases from 2 (SRH recombinations) to 1 for increasing voltage. We rule out trap saturation because it would cause hole recombination to become monomolecular, as the population of trapped electrons becomes constant, and consequently m should increase from 3/2 to 2 ($\alpha_e = \alpha_h = 1$).

3.2.4 Free Energy and Ideality Factor in Perovskite Heterojunctions

The very same method can be readily applied to assess the ideality factor of the single and double heterojunctions that are the building blocks of perovskite solar cells. According to the optical reciprocity relation in Eq. 3.8, the larger the EQY_{PL}, the larger the resulting free energy. Therefore, a purely optical analysis of the EQY_{PL} ought to establish whether additional interface recombination is setting stricter limits to μ_{oc} in single- and double-HP heterojunctions with respect to the bulk. The μ_{oc}–I_{ex} characteristics can be studied much in the same way as the single halide perovskite layers (see Fig. 3.7) and are reported in Fig. 3.9. The μ_{oc}–I_{ex} value deviates from the 3/2 value measured in the intrinsic materials for I_{ex} between 0.01 and 100 suns. Furthermore, the μ_{oc}–I_{ex} characteristics show that m increases with the excitation density, in contrast to what would be expected from the competition between elementary recombination processes. m varies from ≈3/2 to ≈2 in single heterojunctions and from ≈1 to ≈2 in double heterojunctions.

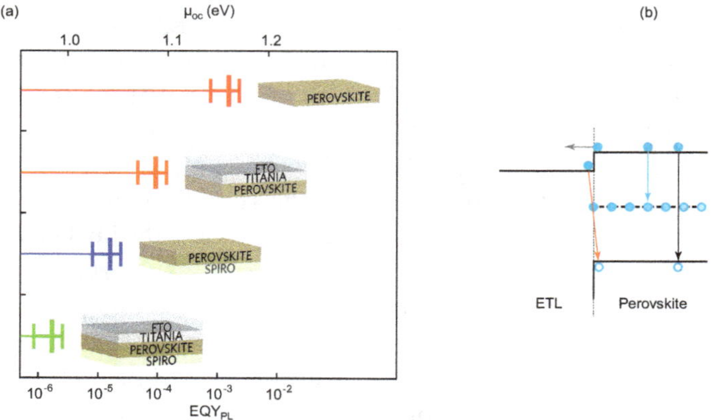

Figure 3.8 Comparison of the free electron–hole energy and external photoluminescence quantum yield in stand-alone perovskite layers and perovskite-based heterojunctions. (a) Top axis: Experimental free energy μ_m *athrmoc* under 50 mW/cm^2 CW laser excitation at 532 nm. Bottom axis: External photoluminescence quantum yield (Z) under the same conditions. The rate of photons absorbed by the film matches that of an illumination level of 1 sun (AM 1.5 G). Two types of single heterojunctions are reported: i-ETL (where the ETL is compact TiO_2) and HTL-i (where the HTL is spiro-MeOTAD). Heterojunctions introduce nonradiative recombination channels, resulting in a lower free energy with respect to the single perovskite layer. (b) Schematic representation of electron–hole recombination processes in a heterojunction: bulk Shockley–Read–Hall decays (cyan arrow), radiative decays (black arrow), and interface decays (orange arrow); the sketch refers to the ETL side interface; a similar one would describe the HTL side.

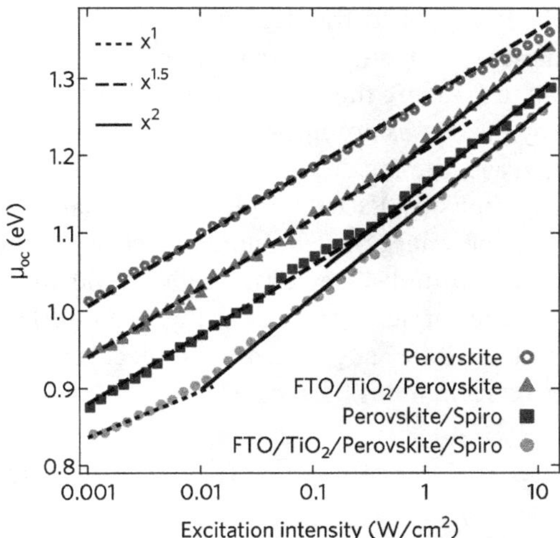

Figure 3.9 μ_{oc}–I_{ex} characteristics and ideality factor of perovskite-based single and double heterojunctions. Markers represent the measured free energy μ_{oc} of electron–hole pairs in perovskite and perovskite-based heterojunctions as a function of the excitation intensity I_{ex} delivered by a CW laser at 532 nm. Lines are provided as a guide to the eye to identify the slope of the data. The ideality factor deviates from the 3/2 value of the single hybrid perovskite layer, increasing to 2 when I_{ex} exceeds a threshold that is peculiar to each type of heterojunction.

Figure 3.8a shows the EQY_{PL} and the corresponding μ_{oc} in these structures at an absorbed monochromatic photon density current corresponding to 1 sun. The stand-alone layer has the largest μ_{oc}, while significantly lower values are measured in the presence of interfaces, in both i-ETL (perovskite-TiO_2) and HTL-i (spiro-MeOTAD-perovskite) heterojunctions. The HTL-i-ETL structure, representing a full-stack but contactless solar cell, shows an even lower EQY_{PL}. This is evidence that additional nonradiative recombination channels appear at both interfaces (Fig. 3.8b), providing faster nonradiative recombination than in the stand-alone layer [52, 53].

Possible interface recombination processes at the heterojunction interfaces are sketched in Fig. 3.8. During steady-state operation, electron (hole) transfer from the i-layer to the ETL (HTL) is compensated by electron (hole) back-transfer, preventing charge build-up on both sides of the interface. We attribute the lower EQY_{PL} observed in our perovskite heterojunctions with respect to the halide perovskite single layers to the activation of such interface decay channels. Quenching of optical excitations at interfaces has even been exploited to measure the diffusion length in halide perovskites. Interface decay processes are not elementary decays, and, as in the case of SRH decays, the ideality factor is expected to increase with the excitation intensity. Excitation-dependent band bending and modification of the energy-level alignment close to the junctions, for example, due to built-in electric fields, could drive nonlinear phenomena for the carrier dynamics at the two interfaces. Anyway, the fact that at 1 sun $m \approx 2$ suggests monomolecular recombinations for both electrons and holes ($\alpha_e \approx \alpha_h \approx 2$).

3.2.5 All-Optical Prediction of the Limit Power Conversion Efficiency

Experimental results provide a detailed assessment of the conversion efficiency of the chemical energy μ into the electrical energy eV, accounting for all recombination losses except those at the electrodes and in the cables. To obtain an explicit estimate of the photoconversion efficiency, the extracted electron (hole) current density can be expressed according to the continuity equation as $J_{e(h)}(\mu) = J_{sun} - J_{rec} = J_{sun} - J_0 \exp(\mu/mkT)$ and the external electrical power as $J_{e(h)}\mu$. J_0 can be experimentally estimated from the open circuit condition $J_{sun} = J_0 \exp(\mu_{oc}/mkT)$. Figure 3.10 shows $eJ_{e(h)}(\mu)$ and the condition for maximum chemical energy conversion, with efficiency $\eta = eJ_{sc} \cdot \mu_{oc} \cdot FF_\mu/I_{sun}$ (FF_μ being the filling factor of the μ-$J_{e(h)}$ characteristics and $eJ_{sc} = eJ_{sun}$). The points in Fig. 3.10 are the experimental data $eJ_{ex} = eJ_{rec}(\mu_{oc})$ from Figs. 3.7 and 3.9 with the substitution $\mu_{oc} \rightarrow \mu$. In the Shockley–Queisser limit, when the solar spectrum is perfectly absorbed down to the bandgap energy, with μ_{rad} = 1.33 eV and m = 1, the ultimate limit η_{SQ} = 30.5% is obtained (eJ_{sc} = 25.4 mA/cm^2, FF_μ = 0.91). In the single halide perovskite layer, SRH limits the chemical energy to μ_{oc} = 1.16 eV and m = 3/2; furthermore, considering the actual absorption of a 250 nm thick halide perovskite film (as typical for solar cells reported in literature), the limiting photoconversion efficiency due to SRH recombination reduces to η_{SRH} = 23.4% (eJ_{sc} = 23.6 mA/cm^2, FF_μ = 0.86). The strictest limit in the investigated double heterojunctions is set by interface recombinations to μ = 0.97 eV ($m \approx 2$), resulting in

η = 18.2% (eJ_{sc} = 23.6 mA/cm^2, FF$_\mu$ = 0.8). Recently, solar cells have been reported based on halide perovskite materials with composite cations, including both organic molecules and the inorganic elements Rb and Cs, with very high external PL efficiency, up to 3.6% [54]. Assuming no electrical losses, the limit efficiency for such cells would be η = 26.2% (d = 400 nm, eJ_{sc} = 24.9 mA/cm^2, μ_{oc} = 1.26 eV, FF$_\mu$ = 0.83).

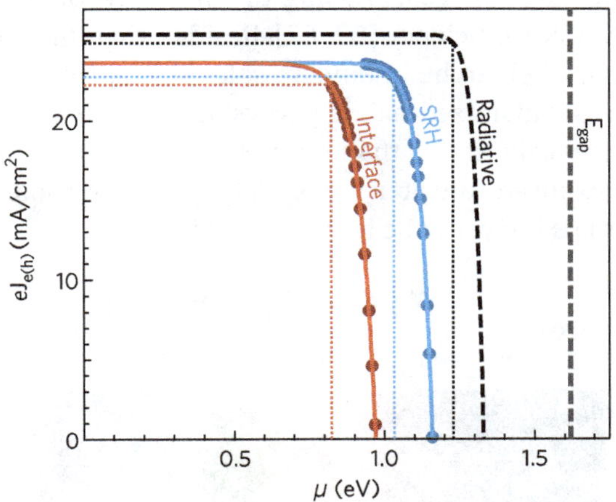

Figure 3.10 Electron (hole) charge current density as a function of the free energy. Solid lines represent the electron (hole) current $J_{e(h)}(\mu) = J_{sun} - J_{rec}$, where J_{sun} is the excitation current at 1 sun (film thickness d = 250 nm) and $J_{rec} = J_0 \exp(\mu/mkT)$ is the diode recombination current in which m and J_0 are the experimental values of the ideality factor and saturation inverse current assessed from the μ-I_{ex} characteristics (Figs. 3.7 and 3.9). Full circles stand for $J_{e(h)}(\mu)$ estimated from the measured diode current $J_{rec}(\mu_{oc})$ with the substitution $\mu_{oc} \rightarrow \mu$. The dotted lines mark the point of maximum extraction of electrical power. Red curve and dots: electron (hole) current density in the double heterojunction; recombination is dominated by interface electron and hole annihilations. Cyan curve and dots: electron (hole) current density in the single hybrid perovskite layer; recombination is due to electron and hole Shockley–Read–Hall annihilations alone. Dashed line: $J_{e(h)}(\mu)$ in the Shockley–Queisser limit.

3.3 Beyond Pure Iodine and Bromine

One attractive feature of halide perovskites is that their bandgaps can be continuously tuned over the whole visible range and up to the near infrared (Fig. 3.11) by exchanging the halide, the metal, the organic group, or a combination of them. Made with relatively earth-abundant materials, mixed-halide perovskites could replace expensive epitaxial semiconductors based on critical raw materials in a variety of applications; these include color-tunable LEDs and laser gain media [7, 33, 35, 36, 55–58]. As photovoltaic absorbers, perovskites of different bandgaps could be employed in multijunction solar cells, raising the Shockley–Queisser limit to power conversion efficiency [59–62]. Besides, their first commercial application is likely to be in tandem solar cells, paired with silicon. In addition to that, mixed-halide perovskites have been proposed to be more chemically stable than pure $MAPbI_3$, which decomposes to PbI_2 when exposed to water vapor [63]. However, some aspects of these materials are yet to be fully understood.

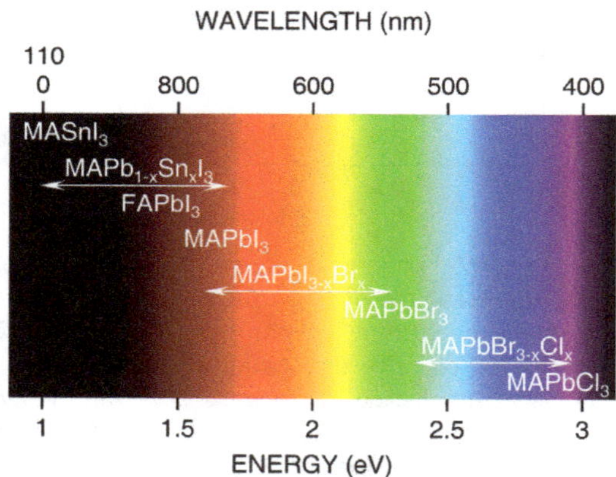

Figure 3.11 Bandgap of halide perovskites in the visible region and near infrared.

The literature on mixed halides is ample, but results are quite varied. Unexpectedly, several reports showed an underwhelming increase—or even a decrease—in V_{oc}, when incorporating the material as the absorber layer in solar cells [64–67]. Hoke et al. [68] looked into the issue by analyzing the optical properties of $MAPbBr_xI_{1-x}$ thin films. They measured the PL of mixed-halide perovskites upon light soaking, under illumination intensities of less than 1 sun. The spectra showed a discrete red-shift to 1.68 eV in less than a minute at room temperature, accompanied by an increase in sub-bandgap absorption. X-ray diffraction patterns showed that the peaks split upon illumination and revert back to their original line shape after a few minutes in the dark. According to their interpretation, this is consistent with reversible light-induced segregation of the mixed-halide alloy, that is halide migration induced by photoexcitation. Similar findings have been reported by various independent works [69–72]. In the case of $MAPb(Br_xI_{1-x})_3$ solar cells, the red-shift in photoluminescence upon light soaking indicates a reduction in the electronic bandgap and quasi-Fermi level splitting, which would explain the lower achievable V_{oc}.

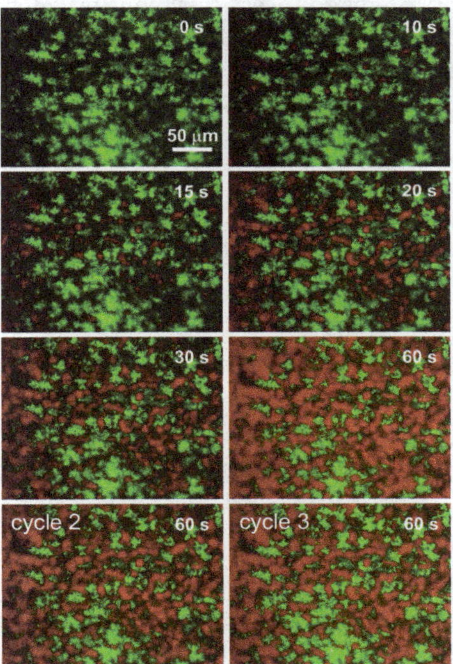

Figure 3.12 Optical microscopy of the photoluminescence of a mixed-halide perovskite film kept under illumination.

The question is still open as to whether phase separation in mixed-halide perovskites could be used to create stable heterostructures. Marongiu et al. [73] demonstrated the creation of self-assembled $MAPbI_{3-x}Br_x$ nanocrystals in a $MAPbBr_3$ matrix. The PL from the films under light soaking has been monitored with an optical microscope equipped with a color camera. Figure 3.12 shows that areas of red emission light up over time and the locations of emission do not change after the sample is cycled through a dark/illumination sequence. This is evidence that iodinerich spots are not dynamically created by halide ion migrations as an effect of light soaking. Rather, optical instabilities are related to the neutralization of traps during illumination, with $MAPbI_{3-x}Br_x$ nanocrystals being funneled optical excitations from the $MAPbBr_3$ matrix, concentrating them in a small fraction of the material's volume and giving rise to optical emission that is efficient even at low excitation intensities. The nanocrystals comprise less than 2% of the materials volume, yet after a photobrightening process they emit brighter luminescence than the matrix, increasing the photoluminescence quantum yield by an order of magnitude.

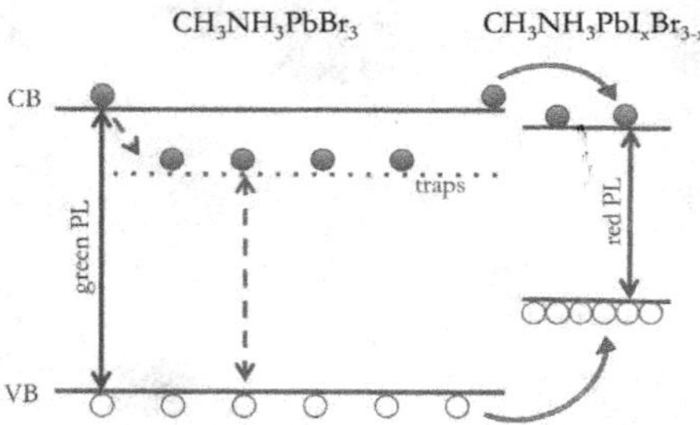

Figure 3.13 Scheme of optical emission in mixed-halide perovskites. Charge transfer from the matrix produces a much larger excitation density in the iodine-rich nanocrystal than in the matrix.

Incorporating delimited regions of charge accumulation could solve the limitations of perovskites as light emitters in a low-injection regime. As explained in Section 3.1.3, radiative recombination in bulk lead halide perovskites is a bimolecular process, so the photoluminescence quantum yield tends to become vanishingly small at low excitation intensities. However, it can grow to the point that it approaches unity when excitation density is close to inversion [33]. As a matter of fact, LEDs based on III–V semiconductors have long been based on double heterojunctions with excited carriers confined in quantum wells, where high photoexcitation densities enhance the probability of radiative recombination between injected electrons and holes. In a similar fashion, iodine-rich nanocrystals embedded in the $MAPbBr_3$ matrix lift the quantum yield by 1 order of magnitude enhancement, thanks to the higher carrier concentration, with the PL intensity being proportional to the product of electron and hole concentrations.

3.4 Conclusions and Outlook

Photophysical evidence points to the fact that excited states of halide perovskites are populated by free electrons and holes, as opposed to excitons [24, 29]. This is a favorable regime when perovskites are used as light absorbers in photovoltaic devices, as an interface is not needed to split the harvested charge carriers. It is a limitation, though, as far as their application as light emitters is concerned. Radiative recombination from an electron–hole plasma is a bimolecular process, with quantum yield approaching unity only at population densities close to inversion. A proposed workaround to achieve high luminescence efficiency at low levels of injection is the engineering of stable nanostructures in a bulk matrix that can confine excited charge carriers into regions of higher concentrations, thus leveraging the higher efficiency [73].

All optical tools show that recombination from the excited state in halide perovskites occurs through the trap-assisted SRH process, which can be identified as the first-order limiting factor to power conversion efficiency in perovskite-based devices [51]. Strategies to approach the ideal Shockley–Queisser limit mostly involve a reduction in the trap density. Replacing a fraction of the organic cations with inorganic cations has been reported to be effective to this purpose [54]. Further losses are introduced in optoelectronic devices by interface recombination at the junction with the transport layers, which might be minimized by engineering better band alignments.

One final point to be argued is that halide perovskites show plenty of untapped potential. Key challenges yet to be tackled include scalability to large-area devices, long-term stability, and low power conversion efficiency in mixed-halide perovskites with the appropriate bandgap for tandem cells [74]. Regardless of any eventual commercial outcome, halide perovskites are likely to fuel a long stretch of further scientific endeavor.

References

1. Stranks, S. D., Eperon, G. E., Grancini, G., Menelaou, C., Alcocer, M. J. P., Leijtens, T., Herz, L. M., Petrozza, A., and Snaith, H. J. (2013). Electron-hole diffusion lengths exceeding 1 micrometer in an organometal trihalide perovskite absorber. *Science*, **342**(6156), 341–344.
2. Wehrenfennig, C., Eperon, G. E., Johnston, M. B., Snaith, H. J., and Herz, L. M. (2013). High charge carrier mobilities and lifetimes in organolead trihalide perovskites. *Adv. Mater.*, **26**(10), 1584–1589.
3. De Wolf, S., Holovsky, J., Moon, S.-J., Löper, P., Niesen, B., Ledinsky, M., Haug, F.-J., Yum, J.-H., and Ballif, C. (2014). Organometallic halide perovskites: sharp optical absorption edge and its relation to photovoltaic performance. *J. Phys. Chem. Lett.*, **5**(6), 1035–1039.
4. Burschka, J., Pellet, N., Moon, S.-J., Humphry-Baker, R., Gao, P., Nazeeruddin, M. K., and Graetzel, M. (2013). Sequential deposition as a route to high-performance perovskite-sensitized solar cells. *Nature*, **499**(7458), 1–5.
5. Baikie, T., Fang, Y., Kadro, J. M., Schreyer, M., Wei, F., Mhaisalkar, S. G., Graetzel, M., and White, T. J. (2013). Synthesis and crystal chemistry of the hybrid perovskite $(CH_3NH_3)PbI_3$ for solid-state sensitised solar cell applications. *J. Mater. Chem. A*, **1**(18), 5628.
6. Buin, A., Pietsch, P., Xu, J., Voznyy, O., Ip, A. H., Comin, R., and Sargent, E. H. (2014). Materials processing routes to trap-free halide perovskites. *Nano Lett.*, **14**(11), 6281–6286.
7. Kojima, A., Teshima, K., Shirai, Y., and Miyasaka, T. (2009). Organometal halide perovskites as visible-light sensitizers for photovoltaic cells. *J. Am. Chem. Soc.*, **131**(17), 6050–6051.
8. Snaith, H. J. (2013). Perovskites: the emergence of a new era for low-cost, high-efficiency solar cells. *J. Phys. Chem. Lett.*, **4**(21), 3623–3630.
9. Green, M. A., Ho-Baillie, A., and Snaith, H. J. (2014). The emergence of perovskite solar cells. *Nat. Photonics*, **8**(7), 506–514.
10. Sutherland, B. R., and Sargent, E. H. (2016). Perovskite photonic sources. *Nat. Photonics*, **10**(5), 295–302.

11. Hirasawa, M., Ishihara, T., Goto, T., Uchida, K., and Miura, N. (1994). Magnetoabsorption of the lowest exciton in perovskite-type compound $(CH_3NH_3)PbI_3$. *Physica B*, **201**, 427–430.

12. Tanaka, K., and Kondo, T. (2003). Bandgap and exciton binding energies in lead-iodide-based natural quantum-well crystals. *Sci. Technol. Adv. Mater.*, **4**(6), 599–604.

13. Tanaka, K., Takahashi, T., Ban, T., Kondo, T., Uchida, K., and Miura, N. (2003). Comparative study on the excitons in lead-halide-based perovskite-type crystals $CH_3NH_3PbBr_3$ $CH_3NH_3PbI_3$. *Solid State Commun.*, **127**(9– 10), 619–623.

14. Lin, Q., Armin, A., Nagiri, R. C. R., Burn, P. L., and Meredith, P. (2015). Electro-optics of perovskite solar cells. *Nat. Photonics*, **9**(2), 106–112.

15. Savenije, T. J., Ponseca Jr., C. S., Kunneman, L., Abdellah, M., Zheng, K., Tian, Y., Zhu, Q., Canton, S. E., Scheblykin, I. G., Pullerits, T., Yartsev, A., and Sundstr¨om, V. (2014). Thermally activated exciton dissociation and recombination control the carrier dynamics in organometal halide perovskite. *J. Phys. Chem. Lett.*, **5**(13), 2189– 2194.

16. Yamada, Y., Nakamura, T., Endo, M., Wakamiya, A., and Kanemitsu, Y. (2014). Photoelectronic responses in solution-processed perovskite $CH_3NH_3PbI_3$ solar cells studied by photoluminescence and photoabsorption spectroscopy. *IEEE J. Photovoltaics*, **5**(1), 401–405.

17. Yan, Y., Yang, M., Choi, S., Zhu, K., Luther, J. M., Yang, Y., and Beard, M. C. (2015). Low surface recombination velocity in solution-grown $CH_3NH_3PbBr_3$ perovskite single crystal. *Nat. Commun.*, **6**, 1–6.

18. Comin, R., Walters, G., Thibau, E. S., Voznyy, O., Lu, Z.-H., and Sargent, E. H. (2015). Structural, optical, and electronic studies of wide-bandgap lead halide perovskites. *J. Mater. Chem. C*, **3**, 8839–8843.

19. Zheng, K., Zhu, Q., Abdellah, M., Messing, M. E., Zhang, W., Generalov, A., Niu, Y., Ribaud, L., Canton, S. E., and Pullerits, T. (2015). Exciton binding energy and the nature of emissive states in organometal halide perovskites. *J. Phys. Chem. Lett.*, **6**(15), 2969–2975.

20. Grancini, G., Kandada, A. R. S., Frost, J. M., Barker, A. J., De Bastiani, M., Gandini, M., Marras, S., Lanzani, G., Walsh, A., and Petrozza, A. (2015). Role of microstructure in the electron-hole interaction of hybrid lead halide perovskites. *Nat. Photonics*, **9**(10), 1–8.

21. Cadelano, M., Saba, M., Sestu, N., Sarritzu, V., Marongiu, D., Chen, F., Piras, R., Quochi, F., Mura, A., and Bongiovanni, G. (2016). Photoexcitations

and emission processes in organometal trihalide perovskites. In Pan, L., and Zhu, G. (eds.), *Perovskite Materials: Synthesis, Characterisation, Properties, and Applications*, InTech, Rijeka.

22. Elliott, R. J. (1957). Intensity of optical absorption by excitons. *Phys. Rev.*, **108**(6), 1384–1389.

23. Sell, D. D., and Lawaetz, P. (1971). New analysis of direct exciton transitions: application to GaP. *Phys. Rev. Lett.*, **26**, 311–314.

24. Sestu, N., Cadelano, M., Sarritzu, V., Chen, F., Marongiu, D., Piras, R., Mainas, M., Quochi, F., Saba, M., Mura, A., and Bongiovanni, G. (2015). Absorption F-sum rule for the exciton binding energy in methylammonium lead halide perovskites. *J. Phys. Chem. Lett.*, **6**(22), 4566–4572.

25. Saha, M. N. (1921). On a physical theory of stellar spectra. *Proc. R. Soc. A*, **99**(697), 135–153.

26. Franciosi, A., Campi, D., Col´ı, G., LaRocca, G. C., Calcagnile, L., Vanzetti, L., DiDio, M., Lomoscolo, M., Cingolani, R., and Rinaldi, R. (1996). Radiative recombination processes in wide-band-gap II–VI quantum wells: the interplay between excitons and free carriers. *JOSA B*, **13**(6), 1268–1277.

27. D'Innocenzo, V., Grancini, G., Alcocer, M. J. P., Kandada, A. R. S., Stranks, S. D., Lee, M. M., Lanzani, G., Snaith, H. J., and Petrozza, A. (2014). Excitons versus free charges in organo-lead tri-halide perovskites. *Nat. Commun.*, **5**, 1–6.

28. Szczytko, J., Kappei, L., Berney, J., Morier-Genoud, F., Portella-Oberli, M. T., and Deveaud, B. (2004). Determination of the exciton formation in quantum wells from time-resolved interband luminescence. *Phys. Rev. Lett.*, **93**(13), 137401–137404.

29. Saba, M., Cadelano, M., Marongiu, D., Chen, F., Sarritzu, V., Sestu, N., Figus, C., Aresti, M., Piras, R., Lehmann, A. G., Cannas, C., Musinu, A., Quochi, F., Mura, A., and Bongiovanni, G. (2014). Correlated electron-hole plasma in organometal perovskites. *Nat. Commun.*, **5**, 5049.

30. Deschler, F., Price, M., Pathak, S., Klintberg, L. E., Jarausch, D.-D., Higler, R., Hu¨ttner, S., Leijtens, T., Stranks, S. D., Snaith, H. J., Atatu¨re, M., Phillips, R. T., and Friend, R. H. (2014). High photoluminescence efficiency and optically pumped lasing in solution-processed mixed halide perovskite semiconductors. *J. Phys. Chem. Lett.*, **5**(8), 1421–1426.

31. Milot, R. L., Eperon, G. E., Snaith, H. J., Johnston, M. B., and Herz, L. M. (2015). Temperature-dependent charge-carrier dynamics in $CH_3NH_3PbI_3$ perovskite thin films. *Adv. Funct. Mater.*, **25**(39), 6218–6227.

32. Wang, H., Whittaker-Brooks, L., and Fleming, G. R. (2015). Exciton and free charge dynamics of methylammonium lead iodide perovskites are different in the tetragonal and orthorhombic phases. *J. Phys. Chem. C*, **119**(34), 19590–19595.

33. Cadelano, M., Sarritzu, V., Sestu, N., Marongiu, D., Chen, F., Piras, R., Corpino, R., Carbonaro, C. M., Quochi, F., Saba, M., Mura, A., and Bongiovanni, G. (2015). Can trihalide lead perovskites support continuous wave lasing?. *Adv. Opt. Mater.*, **3**(11), 1557–1564.

34. Laquai, F. (2014). Materials for lasers: all-round perovskites. *Nat. Mater.*, **13**(5), 429–430.

35. Xing, G., Mathews, N., Lim, S. S., Yantara, N., Liu, X., Sabba, D., Graetzel, M., Mhaisalkar, S., and Sum, T. C. (2014). Low-temperature solution-processed wavelength-tunable perovskites for lasing. *Nat. Mater.*, **13**(5), 476–480.

36. Sarritzu, V., Cadelano, M., Sestu, N., Marongiu, D., Piras, R., Chang, X., Quochi, F., Saba, M., Mura, A., and Bongiovanni, G. (2016). Paving the way for solution-processable perovskite lasers. *Phys. Status Solidi C*, **13**(10–12), 1028–1033.

37. Zhu, H., Fu, Y., Meng, F., Wu, X., Gong, Z., Ding, Q., Gustafsson, M. V., Trinh, M. T., Jin, S., and Zhu, X. (2015). Lead halide perovskite nanowire lasers with low lasing thresholds and high quality factors. *Nat. Mater.*, **14**(6), 636– 642.

38. Fu, Y., Zhu, H., Schrader, A. W., Liang, D., Ding, Q., Joshi, P., Hwang, L., Zhu, X., and Jin, S. (2016). Nanowire lasers of formamidinium lead halide perovskites and their stabilized alloys with improved stability. *Nano Lett.*, **16**(2), 1000–1008.

39. Best research-cell efficiencies. National Renewable Energy Laboratory.

40. Shockley, W., and Queisser, H. J. (1961). Detailed balance limit of efficiency of p-n junction solar cells. *J. Appl. Phys.*, **32**(3), 510–519.

41. Agarwal, S., Seetharaman, M., Kumawat, N. K., Subbiah, A. S., Sarkar, S. K., Kabra, D., Namboothiry, M. A. G., and Nair, P. R. (2014). On the uniqueness of ideality factor and voltage exponent of perovskite-based solar cells. *J. Phys. Chem. Lett.*, **5**(23), 4115–4121.

42. Shi, J., Dong, J., Lv, S., Xu, Y., Zhu, L., Xiao, J., Xu, X., Wu, H., Li, D., Luo, Y., and Meng, Q. (2014). Hole-conductor-free perovskite organic lead iodide heterojunction thin-film solar cells: high efficiency and junction property. *Appl. Phys. Lett.*, **104**(6), 063901–063904.

43. Wetzelaer, Gert-Jan A. H., Scheepers, M., Sempere, A. M., Momblona, C., A′vila, J., and Bolink, H. J. (2015). Trap-assisted non-radiative recombination in organic-inorganic perovskite solar cells. *Adv. Mater.*, **27**(11), 1837–1841.

44. Bi, D., Tress, W., Dar, M. I., Gao, P., Luo, J., Renevier, C., Schenk, K., Abate, A., Giordano, F., Correa-Baena, J. P., Decoppet, J. D., Zakeeruddin, S. M., Nazeeruddin, M. K., Gratzel, M., and Hagfeldt, A. (2016). Efficient luminescent solar cells based on tailored mixed-cation perovskites. *Sci. Adv.*, **2**(1), e1501170–e1501170.

45. Pockett, A., Eperon, G. E., Peltola, T., Snaith, H. J., Walker, A., Peter, L. M., and Cameron, P. J. (2015). Characterization of planar lead halide perovskite solar cells by impedance spectroscopy, open-circuit photovoltage decay, and intensity-modulated photovoltage/photocurrent spectroscopy. *J. Phys. Chem. C*, **119**(7), 3456–3465.

46. Dualeh, A., Moehl, T., T′etreault, N., Teuscher, J., Gao, P., Nazeeruddin, M. K., and Graetzel, M. (2014). Impedance spectroscopic analysis of lead iodide perovskite- sensitized solid-state solar cells. *ACS Nano*, **8**(1), 362–373.

47. Snaith, H. J., Abate, A., Ball, J. M., Eperon, G. E., Leijtens, T., Noel, N. K., Stranks, S. D., Wang, J. T.-W., Wojciechowski, K., and Zhang, W. (2014). Anomalous hysteresis in perovskite solar cells. *J. Phys. Chem. Lett.*, **5**(9), 1511–1515.

48. Unger, E. L., Hoke, E. T., Bailie, C. D., Nguyen, W. H., Bowring, A. R., Heumu¨ller, T., Christoforo, M. G., and McGehee, M. D. (2014). Hysteresis and transient behavior in current–voltage measurements of hybrid-perovskite absorber solar cells. *Energy Environ. Sci.*, **7**(11), 3690–3698.

49. Kim, H.-S., and Park, N.-G. (2014). Parameters affecting I–V hysteresis of $CH_3NH_3PbI_3$ perovskite solar cells: effects of perovskite crystal size and mesoporous TiO_2 layer. *J. Phys. Chem. Lett.*, **5**(17), 2927–2934.

50. Nie, W., Blancon, J.-C., Neukirch, A. J., Appavoo, K., Tsai, H., Chhowalla, M., Alam, M. A., Sfeir, M. Y., Katan, C., Even, J., Tretiak, S., Crochet, J. J., Gupta, G., and Mohite, A. D. (2016). Light-activated photocurrent

degradation and self-healing in perovskite solar cells. *Nat. Commun.*, **7**, 1–9.

51. Sarritzu, V., Sestu, N., Marongiu, D., Chang, X., Masi, S., Rizzo, A., Colella, S., Quochi, F., Saba, M., Mura, A., and Bongiovanni, G. (2017). Optical determination of Shockley-Read-Hall and interface recombination currents in hybrid perovskites. *Sci. Rep.*, 1–10.
52. Marchioro, A., Teuscher, J., Friedrich, D., Kunst, M., van de Krol, R., Moehl, T., Graetzel, M., and Moser, J.-E. (2014). Unravelling the mechanism of photoinduced charge transfer processes in lead iodide perovskite solar cells. *Nat. Photonics*, **8**, 250–255.
53. Ball, J. M., and Petrozza, A. (2016). Defects in perovskite- halides and their effects in solar cells. *Nat. Energy*, **1**(11), 16149.
54. Saliba, M., Matsui, T., Domanski, K., Seo, J. Y., Ummadisingu, A., Zakeeruddin, S. M., Correa-Baena, J. P., Tress, W. R., Abate, A., Hagfeldt, A., and Graetzel, M. (2016). Incorporation of rubidium cations into perovskite solar cells improves photovoltaic performance. *Science*, **354**(6309), 206–209.
55. Jeon, N. J., Noh, J. H., Yang, W. S., Kim, Y. C., Ryu, S., Seo, J., and Il Seok, S. (2015). Compositional engineering of perovskite materials for high-performance solar cells. *Nature*, **517**(7535), 476–480.
56. Tan, Z.-K. (2014). Bright light-emitting diodes based on organometal halide perovskite. *Nat. Nanotechnol.*, **9**(9), 687–692.
57. Sum, T. C., Chen, S., Xing, G., Liu, X., and Wu, B. (2015). Energetics and dynamics in organic–inorganic halide perovskite photovoltaics and light emitters. *Nanotechnology*, **26**(34), 342001–342032.
58. Xing, G., Kumar, M. H., Chong, W. K., Liu, X., Cai, Y., Ding, H., Asta, M., Graetzel, M., Mhaisalkar, S., Mathews, N., and Sum, T. C. (2016). Solution-processed tin-based perovskite for near-infrared lasing. *Adv. Mater.*, 1–6.
59. Gao, P., Graetzel, M., and Nazeeruddin, M. K. (2014). Organohalide lead perovskites for photovoltaic applications. *Energy Environ. Sci.*, **7**(8), 2448–2416.
60. McMeekin, D. P., Sadoughi, G., Rehman, W., Eperon, G. E., Saliba, M., Horantner, M. T., Haghighirad, A., Sakai, N., Korte, L., Rech, B., Johnston, M. B., Herz, L. M., and Snaith, H. J. (2016). A mixed-cation lead mixed-halide perovskite absorber for tandem solar cells. *Science*, **351**(6269), 151–155.

61. Eperon, G. E., Leijtens, T., Bush, K. A., Prasanna, R., Green, T., Wang, J. T. W., McMeekin, D. P., Volonakis, G., Milot, R. L., May, R., Palmstrom, A., Slotcavage, D. J., Belisle, R. A., Patel, J. B., Parrott, E. S., Sutton, R. J., Ma, W., Moghadam, F., Conings, B., Babayigit, A., Boyen, H. G., Bent, S., Giustino, F., Herz, L. M., Johnston, M. B., McGehee, M. D., and Snaith, H. J. (2016). Perovskite-perovskite tandem photovoltaics with optimized band gaps. *Science*, **354**(6314), 861–865.

62. Weidman, M. C., Seltz, M., Stranks, S. D., and Tisdale, W. A. (2016). Highly tunable colloidal perovskite nanoplatelets through variable cation, metal, and halide composition. *ACS Nano*, **10**, 7830–7839.

63. Liu, M., Johnston, M. B., and Snaith, H. J. (2014). Efficient planar heterojunction perovskite solar cells by vapour deposition. *Nature*, **501**(7467), 395–398.

64. Noh, J. H., Im, S. H., Heo, J. H., Mandal, T. N., and Il Seok, S. (2013). Chemical management for colorful, efficient, and stable inorganic–organic hybrid nanostructured solar cells. *Nano Lett.*, **13**(4), 1764–1769.

65. Suarez, B., Gonzalez-Pedro, V., Ripolles, T. S., Sanchez, R. S., Otero, L., and Mora-Sero, I. (2014). Recombination study of combined halides (Cl, Br, I) perovskite solar cells. *J. Phys. Chem. Lett.*, **5**(10), 1628–1635.

66. Kulkarni, S. A., Baikie, T., Boix, P. P., Yantara, N., Mathews, N., and Mhaisalkar, S. (2014). Band-gap tuning of lead halide perovskites using a sequential deposition process. *J. Mater. Chem. A*, **2**(24), 9221–9225.

67. Eperon, G. E., Stranks, S. D., Menelaou, C., Johnston, M. B., Herz, L. M., and Snaith, H. J. (2014). Formamidinium lead trihalide: a broadly tunable perovskite for efficient planar heterojunction solar cells. *Energy Environ. Sci.*, **7**(3), 982–987.

68. Hoke, E. T., Slotcavage, D. J., Dohner, E. R., Bowring, A. R., Karunadasa, H. I., and McGehee, M. D. (2014). Reversible photo-induced trap formation in mixed-halide hybrid perovskites for photovoltaics. *Chem. Sci.*, **6**, 613–617.

69. deQuilettes, D. W., Zhang, W., Burlakov, V. M., Graham, D. J., Leijtens, T., Osherov, A., Bulovicacute, V., Snaith, H. J., Ginger, D. S., and Stranks, S. D. (2016). Photo-induced halide redistribution in organic–inorganic perovskite films. *Nat. Commun.*, **7**, 1–9.

70. Slotcavage, D. J., Karunadasa, H. I., and McGehee, M. D. (2016). Light-induced phase segregation in halide-perovskite absorbers. *ACS Energy Lett.*, **1**(6), 1199–1205.

71. Yoon, S. J., Draguta, S., Manser, J. S., Sharia, O., Schneider, W. F., Kuno, M., and Kamat, P. V. (2016). Tracking iodide and bromide ion segregation in mixed halide lead perovskites during photoirradiation. *ACS Energy Lett.*, **1**(1), 290–296.

72. Galisteo-Lopez, J. F., Li, Y., and M´ıguez, H. (2016). Three-dimensional optical tomography and correlated elemental analysis of hybrid perovskite microstructures: an insight into defect-related lattice distortion and photoinduced ion migration. *J. Phys. Chem. Lett.*, **24**, 5227–5234.

73. Marongiu, D., Chang, X., Sarritzu, V., Sestu, N., Pau, R., Lehmann, A. G., Mattoni, A., Quochi, F., Saba, M., Mura, A., and Bongiovanni, G. (2017). Self-assembled lead halide perovskite nanocrystals in a perovskite matrix. *ACS Energy Lett.*, 769–775.

74. Leo, K. (2015). Perovskite photovoltaics: signs of stability. *Nat. Nanotechnol.*, **10**(7), 574–575.

Chapter 4

Ferroelectricity in Perovskite Solar Cells

Sungkyun Kim, Sang A. Han, Usman Khan, and Sang-Woo Kim
School of Advanced Materials and Engineering, Sungkyunkwan University, Cheoncheon-dong 300, Jangan-gu, Suwon, Gyeonggi-do, 440-746, Korea
kimsw1@skku.edu

This chapter discusses the phenomenon of the efficiency of perovskite solar cells due to the existence of ferroelectricity in perovskite materials. Research on renewable energy sources is very important in securing future energy sources and in environmental aspects. In particular, solar cells are the most industrialized. In recent years, the efficiency of perovskite solar cells, which can replace the high-cost silicon, has dramatically improved. However, the exact cause is still controversial in terms of whether it is due to the hysteresis that occurs in the J–V (current density–voltage) curve or whether it is due to the high power conversion efficiency (PCE) of methylammonium lead triiodide ($MAPbI_3$) perovskite solar cells at the mesoporous electrode. This chapter aims to review that $MAPbI_3$ is a ferroelectric material, by modeling the crystal structure,

Multifunctional Organic–Inorganic Halide Perovskite: Applications in Solar Cells, Light-Emitting Diodes, and Resistive Memory
Edited by Nam-Gyu Park and Hiroshi Segawa

ISBN 978-981-4800-52-5 (Hardcover), 978-1-003-27593-0 (eBook)
www.jennystanford.com

calculating the existence of dipoles, and discussing experimental results using various ferroelectric properties.

4.1 Introduction

Recently, organometal-trihalide perovskite materials, which are widely regarded as materials for solar cells, are emerging as an alternative to silicon-based solar cells because of their low manufacturing cost and high solar cell efficiency [1–5]. Perovskite crystals usually have a chemical composition of ABX_3 [6]. A is a rare earth metal atom, such as Ca, K, Na, Pb, or Sr; and B is a metal cation mainly having a bonding number of 6. X is a halogen atom, such as Cl, Br, or I. Particularly, since perovskite materials have been used as light absorbers to mesoporous solar cells, they have shown increased efficiency and increased interest. Since then, the PCE of the $MAPbI_3$-based solar cell has risen dramatically and has now exceeded 20% efficiency [7].

A dramatic improvement in this performance is the result of the inherent properties of $MAPbI_3$, including favorable direct bandgap, large absorption coefficient, high carrier mobility, and long carrier diffusion length [8–13]. The excellent photovoltaic performance of the $MAPbI_3$ perovskite has been studied in detail, but there is little direct evidence of the fundamental nature of the perovskite. There is a continuing debate among researchers at this point. Hysteresis is also considered to be caused by lattice defects [9, 10], dielectric properties of perovskite [14, 15], and ion migration [16, 17].

In this regard, the ferroelectricity of organometal-trihalide perovskite is verified through comparison with ABO_3 perovskite materials [5, 18]. In general, ferroelectricity was observed with a high dielectric constant in the oxide perovskite material. It also exhibits ferroelectricity in certain phases through phase transitions [10, 19]. Theoretical calculations related to the domain phenomenon of $MAPbI_3$ suggested the possibility of ferroelectricity of $MAPbI_3$ [13, 20, 21]. The ferroelectric domain was also directly observed through piezoresponse measurements. Interestingly, the ferroelectricity of $MAPbI_3$ has been proposed to interpret ionic migration phenomena. Characterization of halide perovskite materials is still underway. In this part, various research methods to prove the ferroelectricity of perovskite are approached and research results are presented.

4.2 Ferroelectric Mechanism in Perovskite

4.2.1 Atomic Modeling

The PCE of hybrid halide perovskite-based solar cells, $MAPbX_3$ (X= Cl, Br, I), has exceeded 20%. This high performance originates from the ferroelectric properties in perovskite materials. It is assumed that this will lower the charge recombination of the electron-hole pair in the solar cell [4, 11, 22].

$MAPbX_3$ is a ferroelectric semiconductor [23, 24] that is different from typical semiconductor materials and ferroelectric materials. The charge transfer mechanism in the perovskite solar cell affects the carrier separation effect due to the polarization and *I–V* (current–voltage) hysteresis [25–28]. In addition, a structural approach can be used to prove that the perovskite is ferroelectric. Since the ferroelectricity of perovskite depends strongly on the atomic structure, it is appropriate to analyze the ferroelectric behavior by evaluating the lattice parameter of $MAPbI_3$. First, the crystal structure parameter of the synthesized $MAPbI_3$ is measured using X-ray diffraction (XRD). As a result of the calculation for all diffraction peaks of the XRD pattern, the lattice parameters can be indexed into I4/mcm space groups of square cells with a = 8.858 Å and c = 12.657 Å. When this $MAPbI_3$ perovskite is represented by a superlattice, one unit cell consists of four ABX_3 units [13, 29, 30].

Therefore, to obtain the lattice parameter of one ABX_3 unit cell, a unit cell containing one ABX_3 constituent unit must be extracted from the superlattice. One ABX_3 unit cell belongs to a square cell in the P4mm space group. The c/a value larger than 1 has already been demonstrated to be capable of spontaneous polarization in the structure of an ABX_3 unit cell like a metal oxide perovskite.

However, the I4/mcm space group is a square center-symmetric group. Tetragonal $SrTiO_3$ perovskite with a I4/mcm space group was reported as a ferroelectric substance. Thus, $MAPbI_3$ perovskite may also be a ferroelectric material.

In this part, we summarize the reports that calculate and quantize the polarization effect of mesoscale ferroelectric materials by modeling the output efficiency of a perovskite solar cell [20, 31–33]. A 3D drift diffusion model has been used to simulate the operation of the solar cell and to incorporate mesoscale ferroelectricity into the perovskite film considering the combined charge at the polarized domain boundary.

Two cases have been simulated to study the relationship between ferroelectricity and low recombination of charge in the device: (i) high recombination strength and (ii) low recombination strength under a ferroelectric polarization scenario. The effect of the polarization direction and the domain size on the output efficiency of the perovskite solar cell was investigated in relation to the charge recombination [34]. To confirm this, structures with different polarizations and domain sizes are modeled. Each structure, for example, a structure in which no polarization exists and a structure polarized in the *X* and *Y* directions, has various polarized domains. The domain structure is calculated by Monte Carlo (MC) simulation using

$$\rho_{\mathrm{pol}} = -\nabla \cdot P \tag{4.1}$$

The combined charge of each structure is calculated in all domains according to Eq. 4.1. Simulated *J–V* characteristics of solar cells for microstructure models were compared in high-recombination-intensity scenarios. The presence of ferroelectric polarization in the region has a significant effect on the solar cell performance at short-circuit current density (J_{SC}) and fill factor (FF) values. It can be seen that the depolarizing electric field resulting from the combined charge at the domain boundary causes carrier accumulation at the domain boundary through calculation of the carrier concentration profile [33, 35].

Domain boundaries can effectively separate charge carriers in a material [36, 37], but the presence of domain boundaries alone does not necessarily account for efficient transport and charge extraction. The dislocation size and length in a solar cell are clearly different from the perpendicular or side of the electrode and the polarization. In the vertical polarization structure, the domain boundary acts as a barrier for efficient carrier transmission. This causes fewer charge carriers to be extracted and recombined at the domain boundaries, resulting in the loss of photocurrent compared to devices with no polarization at the electrode. On the other hand, the charge carriers do not encounter any domain boundaries during transport toward the electrode that extracts high photocurrent in a lateral device structure. Therefore, it can be said that the lateral domain boundary behaves as a similar channel in which separated electrons and holes move along domain boundaries and efficiently transport and induce low recombination loss. In other words, the fill factor (FF), which depends on the interaction between charge transport, extraction, and recombination in the device, is the highest in the lateral structure.

Lateral structure models with small polarization densities exhibit a high FF (84.1%), and it is close to the calculated ideal limit for the FF (88%) of the solar cell that has an open circuit voltage (V_{oc}) = 0.93 V. Therefore, as the polarization density increases further, the ideal limit is reached, and the FF slightly increases.

The vertical device structure has been studied recently in relation to an inorganic ferroelectric photovoltaic cell where the open-circuit voltage is higher than the bandgap of the material. This phenomenon is the result of a high potential created by multiple domain boundaries with potential polarization that can form thick elements (of several micrometers) with strong polarization in some inorganic ferroelectrics.

On the other hand, hybrid perovskite films used in perovskite solar cells are only a few hundred nanometers thick, domain boundaries are small, and hybrid perovskites exhibit low polarization strength [38, 39]. The charge for the domain boundary of the hybrid perovskite films leads to a small potential step at the domain boundary, and the potential does not change significantly along the device. Therefore, no significant improvement in the V_{oc} is observed for the vertical device structure in the perovskite solar cell.

The results for simulations for each of the lateral and vertical devices with respect to changes in domain size are described in terms of *J–V*. To investigate the effect of domain size on perovskite solar cell device performance, we outline the simulations of lateral and vertical device structures with domain sizes of several tens of nanometers, considering high recombination intensities. In addition, the depolarizing field strength resulting from the charge bound to the domain boundary is inversely proportional to the square of the distance from the domain boundary to the domain bulk according to Coulomb's law. This reduces the density of the isolated charge at the domain boundary, which can be efficiently extracted from the electrode. Therefore, a device with a large domain size is inferior in performance.

In the case of a lateral device structure, performance is degraded if the domain size is less than 30 nm. This is due to the recombination of charges separated at the adjacent domain boundaries and increases with a decrease in the domain size. For the vertical device structure, the photocurrent is generated at the anode between the cathode and the domain boundary, where the charge carrier separates at the domain boundary near the electrodes. As the number of domain boundaries decrease the number of domains become smaller and thicker so that there are little photocurrent changes in the vertical devices due to the low polarization. As the polarization of the domain becomes stronger, similar to some inorganic perovskites, a large amount of depolarizing field recombines most of the charge carriers at the domain boundaries, and the photocurrent only occurs in the charges that are located in the first and last domain between the electrodes. This causes the low photocurrent in the inorganic perovskite device. V_{oc} also increases according to the decrease of the domain size and the increase of domain boundaries.

The presence of a highly ordered polarization domain within the perovskite results in reduced charge transport and recombination, and it represents a high FF and J_{SC} [21]. However, such an order will not occur in real devices. In the real devices, random correlated polarizations occur as a high FF because electrons and holes follow the channel of the perovskite, resulting in efficient charge transport and low charge recombination in the solar cell. Finally, the high V_{oc} characteristic of the perovskite solar cell can be accounted for by considering the inherently low recombination strength of the hybrid perovskite.

4.2.2 Crystal Structure with Dipole Moments

A typical perovskite structure has the organic molecule that acts as a coordination cation in a cubic cage made of PbI_3. The stable crystal structure at room temperature (RT) is tetragonal, with transition from 327.4 K to 3D symmetry [40, 41]. Theoretical calculations suggest a small formation energy for MA-I and $MAPbI_3$ vacancies and support the study that the material has a very high defect concentration (10^{17}–10^{20} cm^{-3}) [42, 43]. An interesting argument for high efficiency refers to the ferroelectric properties of perovskite crystals, which can promote charge separation and reduce recombination.

Researchers, initially, expected that strong hysteresis and large permittivity are induced by ferroelectricity [44, 45]. However, a recent study on perovskite shows that hysteretic behavior is induced by mobile ions and charge traps, whereas the macroscopic ferroelectric regions cannot be detected. To illustrate and describe this phenomenon, they have created a crystal model in which all cubic cells are combined with dipole moments considering electrostatic energy. Subsequently, they approximated the cells of the orthotropic system to a cube and assume that the rotation of the polarized molecule will induce a negative charge on the iodide ion.

To find a stable arrangement controlled by the dipole-dipole interaction energy, they operated the Metropolis Monte Carlo (MMC) in a similar way [29, 46]. MMC is a method of obtaining a random sample sequence from a probability distribution that is difficult to sample directly in statistical and statistical physics. This sequence can be used to approximate the histogram generation and calculate the expected value. An important parameter in this method is the dielectric-screening constant of the material. Adjacent dipoles are screened by the typical dielectric response of the semiconductors that have similar bandgaps. The values found in the papers are actually in the range between 5.6 eV and 8.2 eV.

In conclusion, literature suggested MMC and drift-diffusion calculations for steady-state characteristics of perovskite solar cells in the presence of a pattern of directional dipoles. The pattern was obtained by MC operation with energy minimization of dipole-dipole interactions. A short-range antiferroelectric order was found. However, at the 8–10 nm scale, they observed the formation of nanodomains that have a large effect on the static charge of the device. They also induced the formation of nanoconductors through analysis of the parameter space in terms of dipole size and permittivity.

Subsequently, they combined the models obtained in the drift-diffusion simulation, with a focus on the actual role of the conductors in *I–V* characteristics, particularly charge dissociation and recombination losses. This shows that holes and electrons accumulate in the nanodomain and can follow very different current paths to the opposite boundaries. From the above analysis it can be concluded that the dipole can ultimately lead to an increase in photoconversion efficiency due to a reduction in the Shockley–Read–Hall recombination loss and formation of a good current percolation pattern along the nano-dominant edge.

4.2.3 Electronic Structure Calculation

In another aspect, it can be analyzed in terms of materials chemistry and physics of the bulk perovskite as described by the electronic structure calculation [2, 5, 47].

There have been several reports on the structural properties of single crystals of $MAPbI_3$ and related compounds. Polymorphs of upright, square, and cubic were identified, similar to the inorganic perovskites. In the case of $MAPbI_3$, discrepancies between X-ray and electron diffraction are pointed out, suggesting the presence of nanostructured structural domains. Considering the general stoichiometry of ABX_3, charge balance can be achieved in various ways. In the case of the metal oxide perovskite (ABO_3), the total valence of the two metals should be 6.

In the case of halide perovskite, the only possible triplet combination is I-II-X_3, as the valence sum of both cations must be 3. In hybrid halide perovskites, such as $MAPbI_3$, divalent inorganic cations are present and monovalent metals are replaced by organic cations of the same charge. The stability of ionic and heteropolar crystals, such as perovskites, is also influenced by the Madelung electrostatic potential. In the case of the VI anion, the lattice energy decreases as the charge imbalance between A and B is removed. A lower charge is preferred at the site.

In addition, the presence of polar molecules in the center of the perovskite cage has the potential for orientational disorder and polarization. In the case of typical solid-state dielectrics, it has characteristics such as the combination of fast electron (ε_∞) and slow ion polarization contributing to the macroscopic electrostatic dielectric response ($\varepsilon_0 = \varepsilon_\infty + \varepsilon_{ionic} + \varepsilon_{other}$). Molecular reactions (epsilon molecules) can occur for materials that have molecules with permanent dipoles. Such molecular reactions occur more gradually because of the moment of inertia and the dynamic rearrangement of regions.

Organic cations (MA and FA) have the largest and most obvious polarization in the case of methylammonium. The calculation of the polarization tensor in a vacuum using a GAUSSIAN package [13, 48] on a single-charged cation results in the destruction of the crystal center symmetry as the cation of B moves away from the center of BX_6. This result shows the appearance of spontaneous electric polarization and the domination of the molecular polarization tensor by the dipole contribution.

4.3 Ferroelectric Phenomena in the *I–V* Curve

4.3.1 Analysis for Ferroelectric in the Perovskite Solar Cell

Researchers should preferentially test whether this material is included in the 10 ferroelectric material point groups, using XRD data to ensure the existence of the ferroelectric effect [9, 19]. To prove the anisotropic structure of the material, Raman spectroscopy and Fourier-transform infrared spectroscopy (FTIR) spectra can be used. However, because the activities in Raman and infrared (IR) spectroscopy are mutually exclusive in a crystal with inversion symmetry, both Raman and IR spectroscopy can be performed only in the vibrational mode. Finally, the ferroelectric domain structure can be confirmed by polarization–electric field (**P–E**) [49–51] measurement or piezoresponse force microscopy (PFM) [38, 52–55].

4.3.2 Polarization and the Hysteresis Loop

Wei et al. [49] have reported strong bulk polarization of $MAPbI_{3-x}Cl_x$ and $MAPbI_3$ from first-principles density functional theory calculations, and they further applied the energy and electron structure of the domain wall of $MAPbCl_3$, $MAPbI_3$ and $MAPbBr_3$ thin films [56]. The polarization state of the ferroelectric halide light-absorbing material is closely related to the performance of the perovskite solar cell. The hysteresis loop in the *I–V* curve is caused by the ferroelectric effect in the perovskite material. To demonstrate this, various ferroelectricity models are constructed to explain the phenomenon of hysteresis. Assuming that the MA cation in the perovskite can move randomly in the eight directions of the cube, the ferroelectric domain could be present because of the dipole moment of MA. Hence, it is suggested that $CH_3NH_3PbX_3$ (X = I, Br, Cl) is a ferroelectric substance.

The **P–E** curve measurement and the Raman and FTIR spectra were performed to calculate the ferroelectric hysteresis loops [57–59]. The internal electric field of the perovskite solar cell can be illustrated as $\mathbf{E}_{in} = (V_{oc} - V)/W$ within the function of electric field and polarization (V_c is the open-circuit voltage and W the depletion region width) (see Fig. 4.1). The change in polarization occurred after a certain time elapsed in the electric field applied to the ferroelectric domain, and a delay of several seconds occurred in the polarization change even when the electric field became zero. This means that a polarization field is present in the ferroelectric material.

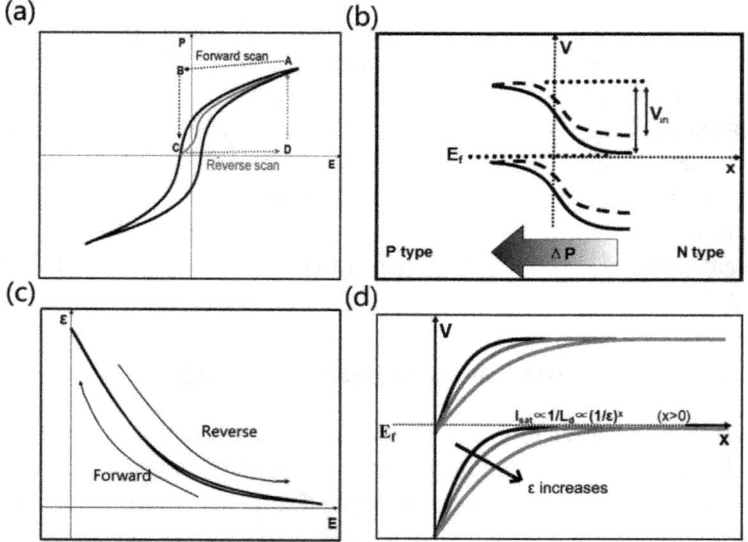

Figure 4.1 (a) Plots; polarization as a function of electric field in ferroelectric materials. (b) Energy band corrected by the excess polarization, induced by the ferroelectric domain in the forward scan mode. (c) Schematic diagram of $\varepsilon(\mathbf{E}) = d\mathbf{P}/d\mathbf{E}$ as a function of electric field in ferroelectric materials. (d) Schematic diagram of the charge region distribution in perovskite. Reprinted with permission from Ref. [49], Copyright 2014, American Chemical Society.

The presence of polarization reduces V_{in} and, consequently, also affects V_{oc}. On the other hand, the short-circuit current I_{sc} is mainly determined by the diffusion length and the transport distance of the charge carrier and thus is not affected by the change in polarization.

Several studies suggest that hysteresis can be observed in the perovskite material at ~0.25 V/s scan rate. This is because the initial polarization condition has a small dielectric constant and cannot follow the scan speed. As the electric field decreases, the polarization changes rapidly in the **P–E** curve. In solar cells, a large dielectric constant change and a small polarization change exhibit higher photocurrent characteristics.

In addition, excluding the ferroelectric effect, a long stepwise measurement was used to determine the photocurrent characteristics of the perovskite [60, 61]. A mesoporous model of the capping layer of a different thickness was used for this purpose [5]. In this step, long, stepwise I–V in forward and reverse modes for each device was conducted. In the forward mode, as the bias voltage increases, the photocurrent initially drops significantly and reaches a steady state of a higher value. Conversely, in the reverse mode, as the voltage decreases, the photocurrent greatly increases and gradually decreases to a steady-state current of a lower value. The dielectric constant of a ferroelectric material changes with the change of the bias, which implies that the existence of a relaxation time in the I–V appears based on the characteristics of the ferroelectric.

The I–V characteristics were determined depending on the dielectric constant of the ferroelectric material by experiments according to the bias voltage. When the dielectric constant increased rapidly at each step, the polarization gradually decreased, and the photocurrent gradually increased. As the voltage step becomes smaller, the dielectric constant decreases to a stable value and the current sharply decreases. In conclusion, hysteresis has also been found to be highly dependent on scan range and scan speed. The ferroelectric effect greatly influences the performance of the hybrid perovskite solar cell according to the structural optimization.

4.3.3 Hysteresis and Ferroelectric Response

Chen et al. [50] showed that there is a strong correlation between the transient ferroelectric polarization of $CH_3NH_3PbI_3$ induced by external bias in the dark condition and hysteresis enhancement. The external electric field affects the direction of the dipole in the perovskite layer and generates polarization, thereby creating an internal electric field (inside the material).

Ferroelectric properties maintain the polarization effect without an applied external bias or phase change of the material. $CH_3NH_3PbI_3$ has a large dielectric constant, of about $10^1 \approx 10^3$, according to the frequency range of 10^{-1}–10^3 Hz. The dielectric constant is further enhanced by external bias and illumination conditions. When a photocurrent is generated in a ferroelectric perovskite, the built-in electric field formed by the photocurrent is exposed to internal polarization because of the dipoles. The optical characteristics of the perovskite devices were found to be strongly influenced by structural differences.

The devices in Fig. 4.2 are mesoporous and planar structures based on TiO_2 and Al_2O_3 materials.

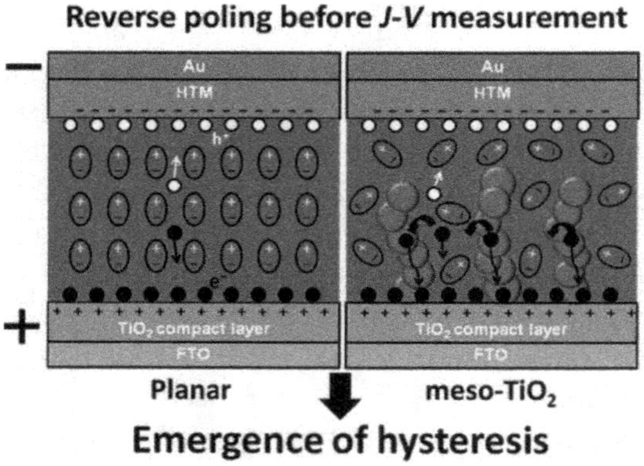

Figure 4.2 Physical mechanism of the polarization in planar and mesoporous perovskite solar cells. Reprinted with permission from Ref. [50], Copyright 2015, American Chemical Society.

There is a large difference in the amount of surface capping layer between the mesoporous and planar structures. In *J*–*V* measurement, the hysteresis behavior of the mesoporous cells depends on the thickness of the capping layer. In addition, the hysteresis of the mesoporous and planar structures was significantly different; Chen et al. [50] showed different V_{oc} along the scan direction, that is, backward scan and forward scan.

All devices were polled by applying an external electric field in the dark condition. After the poling process, the polarization and *J–V* characteristics of $CH_3NH_3PbI_3$ under 1 sun illumination condition were studied. For the negative poling sample, the photocurrent decreased generally in the forward scan, which exhibits a low current density. Photocurrent inhibition also showed a stronger effect on devices with stronger hysteresis. It has been studied that the halogenated perovskite undergoes slow polarization because of ferroelectricity that is sensitive to the applied bias. When the ferroelectricity of a thin film perovskite is sufficiently strong, ordered arrays of dipoles can remain if we do not apply anodic polarization to offset the field. However, when photocurrent is generated, the direction of the electron flow is opposite to the polarization direction. Therefore, it is believed that the remnant polarization restrains the flow of the photocurrent. Aligned dipole arrays can be developed into a planar structure of perovskite. On the other hand, in the mesoporous structure, the ferroelectric domain of the perovskite is weakened by the limited crystal size, and then the dipole may cause random orientation. Most perovskites have a thin capping layer and thus have limited hysteresis. However, if a perovskite has a relatively thick capping layer, internal residual polarization exists in the capping layer and hence hysteresis appears.

Research groups observed that hysteresis tends to increase according to the increased scan rate. This is because dipole arrays cannot follow electric potential change when the scan rate goes high. Moreover, the fact that the poling effect on the *J–V* scan is observed at a low poling potential for planar structures implies that the internal polarization of the perovskite can be very sensitive to the bias conditions prior to the *J–V* measurement.

In addition to reverse poling processing, forward poling of the device in the dark reduces the range of hysteresis. This can be reasonable considering that the internal polarization generated by forward poling has the same direction as the photocurrent. The perovskite cell structure strongly affects the polarization and varies the range of hysteresis prior to *J–V* measurement.

Generally, the ferroelectricity of $CH_3NH_3PbI_3$ is estimated to be weak when it is a bulk crystal. However, ferroelectric polarization can be enhanced when it is in the form of thin film exposed to strong electric field. It is believed that the polarization of the $CH_3NH_3PbI_3$

film is very sensitive to external bias, especially for a continuous planar structure of bulk layer.

In conclusion, it was proved that the ferroelectric polarization of the perovskite is one of the main physical mechanisms that cause *J–V* hysteresis as a result of external bias poling. The stronger hysteresis for the flat heterojunction compared to the mesoporous structure is related to the abundant ferroelectric properties within the bulk crystal in the perovskite. The strong influence of the prebiasing of the cell on the *J–V* characteristic has become clear. These results suggest that ferroelectric polarization inside the cell structure can significantly vary cell performance and hysteresis.

4.4 Piezoresponse in a Perovskite Solar Cell

4.4.1 Study of Piezoresponse and Ferroelectric Domains

In this part, PFM was used to analyze the ferroelectricity of $MAPbI_3$ in a perovskite solar cell. PFM is based on atomic force microscopy (AFM), which is a powerful tool that detects polarization states on the basis of the detection of the local piezoelectric deformation of the ferroelectric sample induced by an external electric field. The driving principle is to measure the converse piezoresponse by generating a local electromechanical oscillation of the sample through the AC voltage applied through a contact between the AFM probe tip and the sample surface. PFM can suffer from artifacts that may occur due to changes in the piezoelectric strain, as well as surface charge, thermal effect, capacitive effects, and electrochemical dipole, and so attention must be paid while data interpretation and analysis [54, 62].

To confirm the ferroelectric property of $MAPbI_3$ in the perovskite solar cells, Kim et al. [38] measured the topography and piezoresponse of $MAPbI_3$ on the fluorine-doped tin oxide (FTO)/glass substrate using AFM and PFM. The topography image of $MAPbI_3$ was built by measuring the *z* axis movement of the AFM tip in contact mode. For the poling process of the $MAPbI_3$ material, they applied DC bias voltage between the conductive AFM probe tip and the FTO layer.

During the PFM measurement, an AC voltage was applied from the conductive tip, as the top electrode. The PFM phase and amplitude of the response of the $MAPbI_3$ material to the electrical signal applied were detected. These effects are often nonlinear, depending on the applied electric field, to induce the second harmonic response. In this study, researchers have developed a customized second harmonic PFM technique to investigate possible nonlinear effects in an apparent piezoelectric response and have studied various electromechanical mechanisms.

To understand the origin of the piezoelectric properties in $MAPbI_3$, PFM was measured by applying an AC bias using a lock-in device (Fig. 4.3). As a result of measurement of PFM amplitude and PFM phase, polarization occurred due to spontaneous polarization inside $MAPbI_3$. The intensity of PFM amplitude in the perovskite crystals was significantly different depending on the crystal size. Also, it was confirmed that most of the dipoles inside the perovskite material are aligned in the same direction through the PFM phase.

In the absence of an electric field, it showed a slight phase difference in $MAPbI_3$ and spontaneous polarization and when an electric field was applied, the phase difference was more prominent. Quantitative changes in the phase angle and amplitude occur under external bias. The polarization behavior of $MAPbI_3$ seems to be due to the large free rotation of the polar M cation, unlike the change in the displacement of the B site cation of the perovskite oxide. When a positive bias is applied to the surface, it is clearly observed that the PFM phase changes to +90°. Conversely, a negative bias indicates that the direction of the aligned dipole is reversed. The phase inversion did not occur in all regions, but the higher the crystallinity, the more noticeable the PFM phase difference. The polarization transition of $MAPbI_3$ is much lower than that of other inorganic ferroelectric thin films. Nonetheless, $MAPbI_3$ has proven to have switching domains and ferroelectricity exhibiting polarization hysteresis. At the domain boundary, the phase is inverted, which establishes a linear piezoelectric property of $MAPbI_3$ and the 180° PFM phase difference is expected to be caused by the spontaneous polarization of $MAPbI_3$, possibly aligned antiparallel to each other.

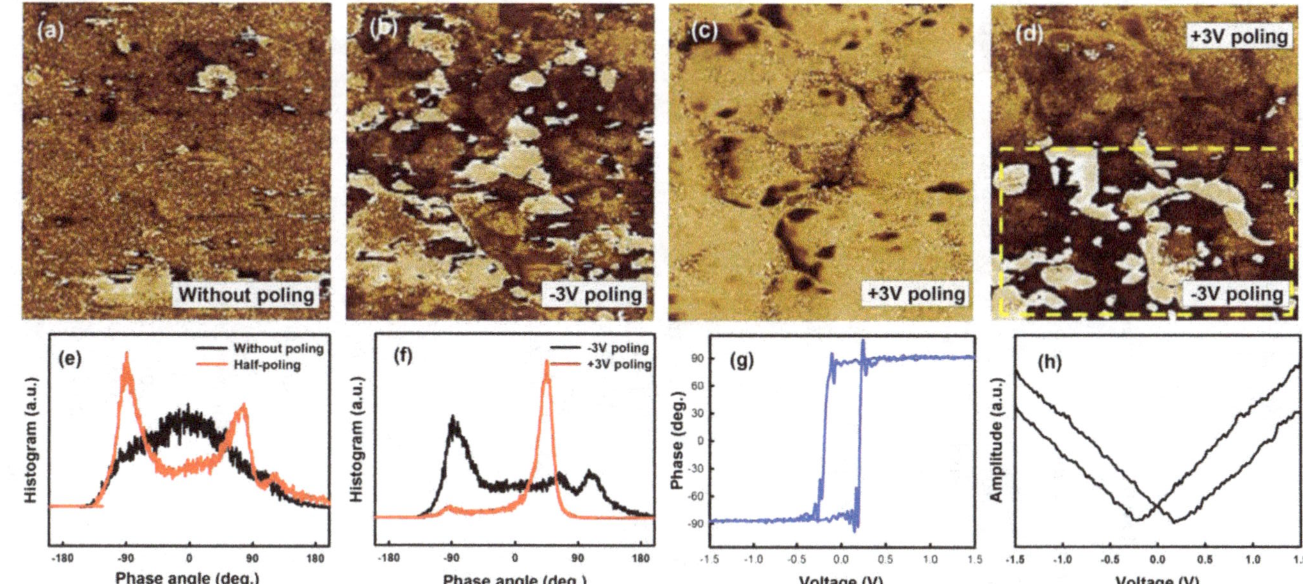

Figure 4.3 PFM phase for a $MAPbI_3$ thin film under different poling conditions: (a) Without poling, (b) –3 V poling, (c) +3 V poling, and (d, top) +3 V poling and (d, bottom) –3 V poling (e) Domain phase distribution and (f) Domain phase distribution for poling (g) Piezoresponse phase and (h) piezoresponse amplitude. Reprinted with permission from Ref. [54], Copyright 2015, Elsevier.

4.4.2 Study of Surface Potential and Photoinduced Shifting

The spontaneous polarization of perovskite materials will induce an internal electric field, which will attract charge to the surface and affect the surface potential. Kelvin probe force microscopy (KPFM) can measure the surface potential of a sample and analyze the surface charge density (Fig. 4.4) [62–64].

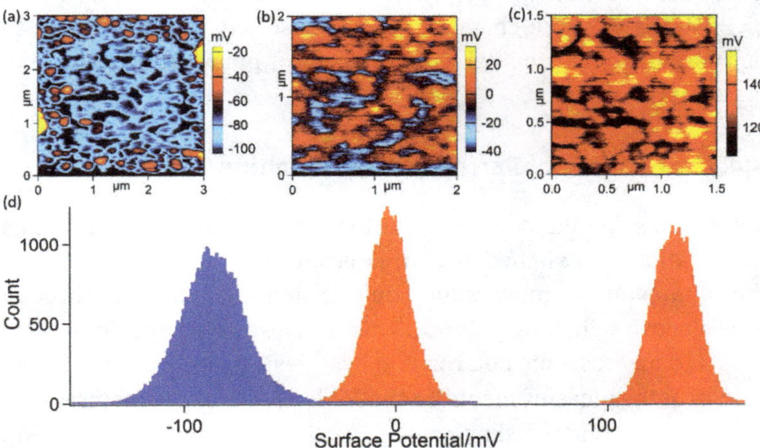

Figure 4.4 Surface potential mapping of a $MAPbI_3$ film by KPFM, with a region of (a) negative, (b) near-zero, and (c) positive surface potential. (d) Histogram of surface potential mappings in a $MAPbI_3$ film. Reprinted with permission from Ref. [62], Copyright 2009, Royal Society of Chemistry.

Surface potential is so related to the work function of the material. However, in the case of a ferroelectric, the direction of the internal polarization is controlled to cause an unstable state of the surface. To stabilize surface potential, external charges adhere to the sample surface to create a difference in surface potential.

The polarization direction and the flow of electrons in the perovskite material can promote or impede charge transport. In addition, light may pass through the $MAPbI_3$ thin films. When $MAPbI_3$ is irradiated with light, the surface potential is shifted. In this configuration, electrons move down to the TiO_2 while moving up to the surface until the holes become equilibrium.

To further determine whether the observed surface potential shift was actually due to the photoinduced charge, the change in the surface potential distribution of the $MAPbI_3$ film on the FTO/poly(3,4-ethylenedioxythiophene):poly(styrene sulfonate) substrate was measured for several cycles of the light turning on and off. The surface potential is clearly shifted toward a positive value when the light is turned on and falls off again when the light is off, and the trend continues as the cycle repeats. A similar observation of the photoinduced surface potential shift is confirmed by different samples. In experiments with continuous light, the surface potential increased with time and returned to its original position. As a result of PFM analysis, the change of PFM phase and amplitude due to light occurred, but the persistence effect of polarization inside the material was short.

4.4.3 Analysis of Ferroelectric Switching

$MAPbI_3$ has spontaneous polarization, and it is confirmed that this polarization is switched by an external electric field. Therefore, the photovoltaic conversion could potentially change through ferroelectric switching [51, 65]. This is investigated by microscopic PFM and macroscopic polarization hysteresis measurements. In the results of the measurements on the $MAPbI_3$ film, the hysteresis loop of the voltage and the shape of the amplitude-voltage butterfly loop are obtained, and this loop is generally interpreted as a signature for ferroelectric switching. The ferroelectric switching characteristics are evident in the *J–V* curve along the poling direction [62]. Additionally, the light illumination reduces the energy barrier for switching, thereby reducing the coercive voltage. It also suggests moving the coercive voltage by inducing an internal bias that prefers different deflection directions.

Ferroelectric switching and photovoltaic effects in $MAPbI_3$ are different from conventional perovskite materials (Fig. 4.5). In addition, antivoltage reduction in the piezoelectric response can be caused by light-induced ion movement. Migration of photoinduced halogen compounds away from the illuminated region has been reported, which can lead to a reduction in ferroelectric polarization as the first principles calculations reveal that the PbI_3 anions are important contributors to polarization. This reduced polarization can reduce the breakdown voltage and piezoelectric response, and the coercive voltage can be further reduced by increased energy from the light to overcome the switching barrier.

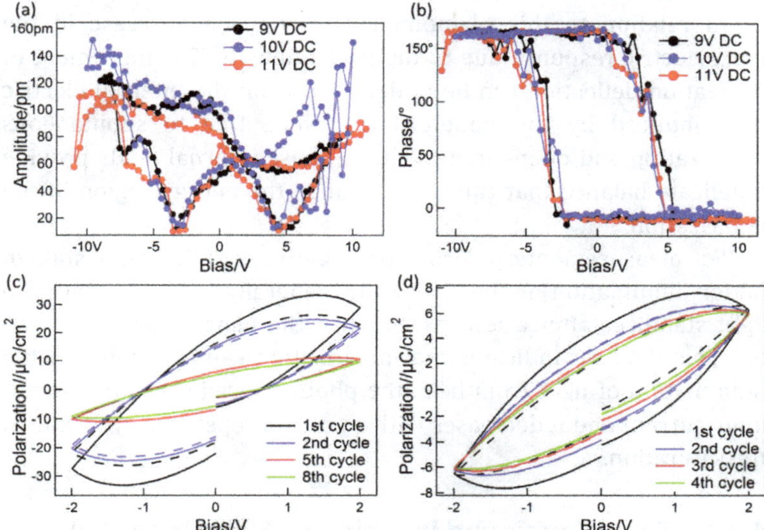

Figure 4.5 Ferroelectric switching phenomena in $MAPbI_3$. (a, b) PFM amplitude and phase depending on the bias voltage. (c, d) Hysteresis loops at different cycles switched under dark and under illumination. Reprinted with permission from Ref. [62], Copyright 2009, Royal Society of Chemistry.

The evolution of ferroelectric polarization during the cycle is clearly undesirable, which can affect the efficiency stability of the device. The ferroelectric domain structure needs to be stabilized against both electric field and light irradiation. In addition, the depolarization field can enhance or reduce photocell performance by promoting or hindering the transfer of photoinduced charge. This means that proper poling may be effective in improving photovoltaic properties, but ion motion can be triggered, and the process must be carefully designed.

That the spontaneous polarization can be converted by illumination even in the absence of an electric field demonstrates a strong interaction between polarization and light in technically feasible $MAPbI_3$ films. Light is displayed to reduce the coercive voltage of $MAPbI_3$ and shifts the nucleation biases, indicating that photoelectric conversion is caused by ionic motion in combination with photovoltaic cells. This series of studies provides strong evidence for the interaction between photoinduced charge, polarization, and ion in perovskite $MAPbI_3$.

In relation to this phenomenon, there is a decrease in the piezoelectric response due to the light intensity. The movement of nucleation deflection can be understood from the internal electric field induced by the photoelectric effect. Due to spontaneous polarization and open-circuit voltage, these internal fields provide a delicate balance that can easily change the energy region of the polarization.

For measurements, polarization, regardless of the initial state of under illumination or the dark state, generally increases until the light stabilizes after a few cycles when the light is turned on and off. This data set indicates that polarization can occur due to the illumination of light from both the photoinduced electron and the ionic current and it decreases with a gradual repetition cycle due to the saturation.

4.4.4 Ferroelectric and Pyroelectric Effect in $MAPbI_3$

A study on the pyroelectric property of $MAPbI_3$ has been carried out (Fig. 4.6) [21, 66, 67]. The periodic pulse of the pyroelectric output was measured in order to determine whether the square $MAPbI_3$ had polarity along the *c* axis. To distinguish the effect of the other thermally stimulated electric response (TSER), the superconductor was heated once and then heat was applied to another part of the crystal. Pyroelectric currents will reverse their sign once a crystal, together with their electrical leads, is flipped. However, thermoelectricity is created by the temperature gradient between the conductors or semiconductors and does not depend on the internal orientation of the dipole.

It shows a clear response from the superconductor when the electrode is deposited on a <001> plane. Since the conductivity of a semiconductor increases with temperature, it means that the thermoelectric current can be dominant at higher temperatures. The experiment is performed at a low temperature of ~210 K to reduce the thermoelectric contribution to the TSER. At RT, the superconducting response still exists but is much weaker than at low temperatures, where TSER is dominant.

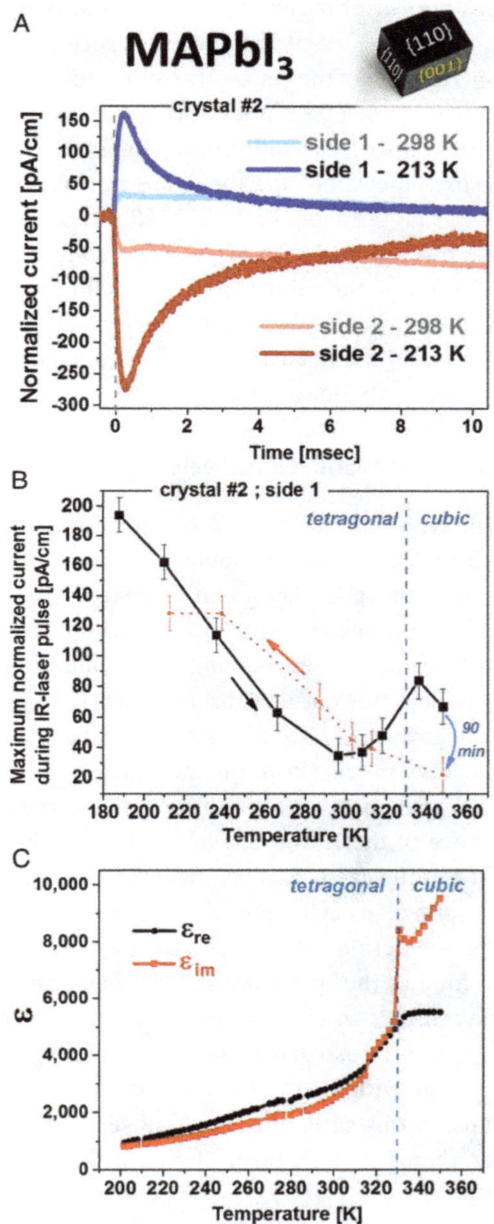

Figure 4.6 Pyroelectric response and relative permittivity measurements. Reprinted with permission from Ref. [66].

Another evidence of the pyroelectric nature of $MAPbI_3$ is presented that is based on peak current value dependence on temperature. The maximum value near the phase transition temperature means an increase of the pyroelectric response in the ferroelectric-phase dielectric phase transition based on the Ginsburg–Landau theory. Therefore, the extreme value near the phase transition temperature is clear evidence that tetragonal crystal $MAPbI_3$ is ferroelectric. The disappearance and reproduction of the superconducting response after exceeding the T_c in the cubic phase and cooling to the tetragonal crystallization step provides further support for the polarity of the tetragonal crystal and the polarity of the tetragonal crystal $MAPbI_3$, which is consistent with the results.

4.4.5 Second Harmonic Generation

If the Rayleigh range of the excited beam is much smaller than the crystal thickness, second harmonic generation (SHG) can only be observed near the microcrystalline surface due to the phase [23, 66]. SHG is a nonlinear optical process coupled with photons with the same frequency interacting with nonlinear materials. Experimentally, the stripe pattern observed in SHG can be caused by the ferroelectric domain of $MAPbI_3$ visible after the crystal is etched.

Regardless of the precise location, the signal for the polarization similar to the bipolar pattern over the measured region coincides with the presence of the ferroelectric domain. This is because the orientation of synthesized $MAPbI_3$, which determines the SHG polarization response in each region, is generally similar.

Unlike papers where no evidence of SHG activity in the existing $MAPbI_3$ was found, the presence of ferroelectric domains in polycrystalline samples was confirmed. When multiple ferroelectric domains are present, destructive interference between scattered SHGs from domains with inverted polarity is likely to result in very low SHG signals. In this case, in order to observe the SHG, a single domain larger than the scan pixel of the SHG must be optically excited.

4.5 Conclusion

In summary, we have provided the ferroelectric nature of $MAPbI_3$ and confirmed that the ferroelectric polarization of the material improves performance of the organometallic halide perovskite solar cell.

First, we explained that $MAPbI_3$ is ferroelectric through simulation and modeling calculations: the crystal structure through atomic modeling is similar to the well-known oxide perovskite materials, hysteresis, and *J*–*V* curves in the crystal structure with dipole moment. In addition, various experimental results have shown that the tetragonal crystals of $MAPbI_3$ are ferroelectrics. The presence of domains in the material through PFM and KPFM and the polarization reversal phenomenon under external electric fields; the absence of inverse symmetry demonstrated by the presence of SHG; and the spontaneous polarization and polarization domains through pyroelectricity. In addition, ferroelectric polarization can control the interfacial band structure of the perovskite solar cell. By the polarization within the perovskite solar cell, the forward scan of the *J*–*V* curve leads to lower PCE because of the negative polarity region. On the other hand, higher efficiency can be achieved by the reverse scan because of the anode region. Electric poling and scanning directions have a significant effect on the photovoltaic performance of solar cells and are common in ferroelectric-based solar cells.

In the future, convergence studies based on in-depth analysis of the device structure and the experimental phenomena should be conducted in various fields, such as crystal structure, lattice vibration, and charge mobility, as well as ferroelectricity. To achieve this, more studies will be necessary to understand the new perovskite materials, stable device driving, and efficiency improvements.

References

1. Yang, W. S., Noh, J. H., Jeon, N. J., Kim, Y. C., Ryu, S., Seo, J., and Seok, S. I. (2015). High-performance photovoltaic perovskite layers fabricated through intramolecular exchange. *Science*, **348**, 1234–1237.

2. Zhou, H. P., Chen, Q., Li, G., Luo, S., Song, T. B., Duan, H. S., Hong, Z. R., You, J. B., Liu, Y. S., and Yang, Y. (2014). Interface engineering of highly efficient perovskite solar cells. *Science*, **345**, 542–546.
3. Kojima, A., Teshima, K., Shirai, Y., and Miyasaka, T. (2009). Organometal halide perovskites as visible-light sensitizers for photovoltaic cells. *J. Am. Chem. Soc.*, **131**, 6050.
4. Kim, H. S., Lee, C. R., Im, J. H., Lee, K. B., Moehl, T., Marchioro, A., Moon, S. J., Humphry-Baker, R., Yum, J. H., Moser, J. E., Gratzel, M., and Park, N. G. (2012). Lead iodide perovskite sensitized all-solid-state submicron thin film mesoscopic solar cell with efficiency exceeding 9%. *Sci. Rep.*, **2**.
5. Lee, M. M., Teuscher, J., Miyasaka, T., Murakami, T. N., and Snaith, H. J. (2012). efficient hybrid solar cells based on meso-superstructured organometal halide perovskites. *Science*, **338**, 643–647.
6. Liu, M. Z., Johnston, M. B., and Snaith, H. J. (2013). Efficient planar heterojunction perovskite solar cells by vapour deposition. *Nature*, **501**, 395.
7. Park, N. G. (2013). Organometal perovskite light absorbers toward a 20% efficiency low-cost solid-state mesoscopic solar cell. *J. Phys. Chem. Lett.*, **4**, 2423–2429.
8. Burschka, J., Pellet, N., Moon, S. J., Humphry-Baker, R., Gao, P., Nazeeruddin, M. K., and Gratzel, M. (2013). Sequential deposition as a route to high-performance perovskite-sensitized solar cells. *Nature*, **499**, 316.
9. Jeon, N. J., Noh, J. H., Kim, Y. C., Yang, W. S., Ryu, S., and Seok, S. I. (2014). Solvent engineering for high-performance inorganic-organic hybrid perovskite solar cells. *Nat. Mater.*, **13**, 897–903.
10. Jeon, N. J., Noh, J. H., Yang, W. S., Kim, Y. C., Ryu, S., Seo, J., and Seok, S. I. (2015). Compositional engineering of perovskite materials for high-performance solar cells. *Nature*, **517**, 476.
11. Nie, W. Y., Tsai, H. H., Asadpour, R., Blancon, J. C., Neukirch, A. J., Gupta, G., Crochet, J. J., Chhowalla, M., Tretiak, S., Alam, M. A., Wang, H. L., and Mohite, A. D. (2015). High-efficiency solution-processed perovskite solar cells with millimeter-scale grains. *Science*, **347**, 522–525.
12. Zhao, P., Yin, W., Kim, M., Han, M., Song, Y. J., Ahn, T. K., and Jung, H. S. (2017). Improved carriers injection capacity in perovskite solar cells by introducing A-site interstitial defects. *J. Mater. Chem. A*, **5**, 7905–7911.

13. Frost, J. M., Butler, K. T., Brivio, F., Hendon, C. H., van Schilfgaarde, M., and Walsh, A. (2014). Atomistic origins of high-performance in hybrid halide perovskite solar cells. *Nano Lett.*, **14**, 2584–2590.

14. Green, M. A., Ho-Baillie, A., and Snaith, H. J. (2014). The emergence of perovskite solar cells. *Nat. Photonics*, **8**, 506–514.

15. Juarez-Perez, E. J., Sanchez, R. S., Badia, L., Garcia-Belmonte, G., Kang, Y. S., Mora-Sero, I., and Bisquert, J. (2014). Photoinduced giant dielectric constant in lead halide perovskite solar cells. *J. Phys. Chem. Lett.*, **5**, 2390–2394.

16. Xiao, Z. G., Yuan, Y. B., Shao, Y. C., Wang, Q., Dong, Q. F., Bi, C., Sharma, P., Gruverman, A., and Huang, J. S. (2015). Giant switchable photovoltaic effect in organometal trihalide perovskite devices (vol. 14, 193, 2014). *Nat. Mater.*, **14**.

17. Gottesman, R., Haltzi, E., Gouda, L., Tirosh, S., Bouhadana, Y., and Zaban, A. (2014). Extremely slow photoconductivity response of $CH_3NH_3PbI_3$ perovskites suggesting structural changes under working conditions. *J. Phys. Chem. Lett.*, **5**, 2662–2669.

18. Nuraje, N., and Su, K. (2013). Perovskite ferroelectric nanomaterials. *Nanoscale*, **5**, 8752–8780.

19. Baikie, T., Fang, Y. N., Kadro, J. M., Schreyer, M., Wei, F. X., Mhaisalkar, S. G., Graetzel, M., and White, T. J. (2013). Synthesis and crystal chemistry of the hybrid perovskite (CH_3NH_3) PbI_3 for solid-state sensitised solar cell applications. *J. Mater. Chem. A*, **1**, 5628–5641.

20. Liu, S., Zheng, F., Koocher, N. Z., Takenaka, H., Wang, F. G., and Rappe, A. M. (2015). Ferroelectric domain wall induced band gap reduction and charge separation in organometal halide perovskites. *J. Phys. Chem. Lett.*, **6**, 693–699.

21. Butler, K. T., Frost, J. M., and Walsh, A. (2015). Ferroelectric materials for solar energy conversion: photoferroics revisited. *Energy Environ. Sci.*, **8**, 838–848.

22. Ahn, N., Son, D. Y., Jang, I. H., Kang, S. M., Choi, M., and Park, N. G. (2015). Highly reproducible perovskite solar cells with average efficiency of 18.3% and best efficiency of 19.7% fabricated via lewis base adduct of lead(II) iodide. *J. Am. Chem. Soc.*, **137**, 8696–8699.

23. Liao, W. Q., Zhang, Y., Hu, C. L., Mao, J. G., Ye, H. Y., Li, P. F., Huang, S. P. D., and Xiong, R. G. (2015). A lead-halide perovskite molecular ferroelectric semiconductor. *Nat. Commun.*, **6**.

24. Snaith, H. J., Abate, A., Ball, J. M., Eperon, G. E., Leijtens, T., Noel, N. K., Stranks, S. D., Wang, J. T. W., Wojciechowski, K., and Zhang, W. (2014).

Anomalous hysteresis in perovskite solar cells. *J. Phys. Chem. Lett.*, **5**, 1511–1515.

25. Frost, J. M., Butler, K. T., and Walsh, A. (2014). Molecular ferroelectric contributions to anomalous hysteresis in hybrid perovskite solar cells. *APL Mater.*, **2**, 081506.

26. van Reenen, S., Kemerink, M., and Snaith, H. J. (2015). Modeling anomalous hysteresis in perovskite solar cells. *J. Phys. Chem. Lett.*, **6**, 3808–3814.

27. Zhang, T., Chen, H. N., Bai, Y., Xiao, S., Zhu, L., Hu, C., Xue, Q. Z., and Yang, S. H. (2016). Understanding the relationship between ion migration and the anomalous hysteresis in high-efficiency perovskite solar cells: a fresh perspective from halide substitution. *Nano Energy*, **26**, 620–630.

28. Zhang, Y., Liu, M. Z., Eperon, G. E., Leijtens, T. C., McMeekin, D., Saliba, M., Zhang, W., de Bastiani, M., Petrozza, A., Herz, L. M., Johnston, M. B., Lin, H., and Snaith, H. J. (2015). Charge selective contacts, mobile ions and anomalous hysteresis in organic-inorganic perovskite solar cells. *Mater. Horiz.*, **2**, 315–322.

29. Pecchia, A., Gentilini, D., Rossi, D., Auf der Maur, M., and Di Carlo, A. (2016). Role of ferroelectric nanodomains in the transport properties of perovskite solar cells. *Nano Lett.*, **16**, 988–992.

30. Noh, J. H., Im, S. H., Heo, J. H., Mandal, T. N., and Seok, S. I. (2013). Chemical management for colorful, efficient, and stable inorganic-organic hybrid nanostructured solar cells. *Nano Lett.*, **13**, 1764–1769.

31. Bi, F. Z., Markov, S., Wang, R. L., Kwok, Y. H., Zhou, W. J., Liu, L. M., Zheng, X., Chen, G. H., and Yam, C. Y. (2017). Enhanced photovoltaic properties induced by ferroelectric domain structures in organometallic halide perovskites. *J. Phys. Chem. C*, **121**, 11151–11158.

32. Bai, S., Wu, Z. W., Wu, X. J., Jin, Y. Z., Zhao, N., Chen, Z. H., Mei, Q. Q., Wang, X. Z., Ye, Z., Song, T. Y., Liu, R. Y., Lee, S. T., and Sun, B. Q. (2014). High-performance planar heterojunction perovskite solar cells: preserving long charge carrier diffusion lengths and interfacial engineering. *Nano Res.*, **7**, 1749–1758.

33. Sherkar, T. S., and Koster, L. J. A. (2016). Can ferroelectric polarization explain the high performance of hybrid halide perovskite solar cells?. *Phys. Chem. Chem. Phys.*, **18**, 331–338.

34. Ponseca, C. S., Savenije, T. J., Abdellah, M., Zheng, K. B., Yartsev, A., Pascher, T., Harlang, T., Chabera, P., Pullerits, T., Stepanov, A., Wolf, J. P., and Sundstrom, V. (2014). Organometal halide perovskite solar

cell materials rationalized: ultrafast charge generation, high and microsecond-long balanced mobilities, and slow recombination. *J. Am. Chem. Soc.*, **136**, 5189–5192.

35. Genenko, Y. A., Glaum, J., Hoffmann, M. J., and Albe, K. (2015). Mechanisms of aging and fatigue in ferroelectrics. *Mater. Sci. Eng. B*, **192**, 52–82.

36. Rohm, H., Leonhard, T., Hoffmannbc, M. J., and Colsmann, A. (2017). Ferroelectric domains in methylammonium lead iodide perovskite thin-films. *Energy Environ. Sci.*, **10**, 950–955.

37. Rothmann, M. U., Li, W., Zhu, Y., Bach, U., Spiccia, L., Etheridge, J., and Cheng, Y. B. (2017). Direct observation of intrinsic twin domains in tetragonal $CH_3NH_3PbI_3$. *Nat. Commun.*, **8**.

38. Kim, H. S., Kim, S. K., Kim, B. J., Shin, K. S., Gupta, M. K., Jung, H. S., Kim, S. W., and Park, N. G. (2015). Ferroelectric polarization in $CH_3NH_3PbI_3$ perovskite. *J. Phys. Chem. Lett.*, **6**, 1729–1735.

39. Almora, O., Zarazua, I., Mas-Marza, E., Mora-Sero, I., Bisquert, J., and Garcia-Belmonte, G. (2015). Capacitive dark currents, hysteresis, and electrode polarization in lead halide perovskite solar cells. *J. Phys. Chem. Lett.*, **6**, 1645–1652.

40. Wang, Q., Lyu, M. Q., Zhang, M., Yun, J. H., Chen, H. J., and Wang, L. Z. (2015). Transition from the tetragonal to cubic phase of organohalide perovskite: the role of chlorine in crystal formation of $CH_3NH_3PbI_3$ on TiO_2 substrates. *J. Phys. Chem. Lett.*, **6**, 4379–4384.

41. Quarti, C., Mosconi, E., Ball, J. M., D'Innocenzo, V., Tao, C., Pathak, S., Snaith, H. J., Petrozza, A., and De Angelis, F. (2016). Structural and optical properties of methylammonium lead iodide across the tetragonal to cubic phase transition: implications for perovskite solar cells. *Energy Environ. Sci.*, **9**, 155–163.

42. Zheng, X. P., Chen, B., Dai, J., Fang, Y. J., Bai, Y., Lin, Y. Z., Wei, H. T., Zeng, X. C., and Huang, J. S. (2017). Defect passivation in hybrid perovskite solar cells using quaternary ammonium halide anions and cations. *Nat. Energy*, **2**.

43. Mei, A. Y., Li, X., Liu, L. F., Ku, Z. L., Liu, T. F., Rong, Y. G., Xu, M., Hu, M., Chen, J. Z., Yang, Y., Gratzel, M., and Han, H. W. (2014). A hole-conductor-free, fully printable mesoscopic perovskite solar cell with high stability. *Science*, **345**, 295–298.

44. Tress, W., Marinova, N., Moehl, T., Zakeeruddin, S. M., Nazeeruddin, M. K., and Gratzel, M. (2015). Understanding the rate-dependent J-V hysteresis, slow time component, and aging in $CH_3NH_3PbI_3$ perovskite

solar cells: the role of a compensated electric field. *Energy Environ. Sci.*, **8**, 995–1004.

45. Unger, E. L., Hoke, E. T., Bailie, C. D., Nguyen, W. H., Bowring, A. R., Heumuller, T., Christoforo, M. G., and McGehee, M. D. (2014). Hysteresis and transient behavior in current-voltage measurements of hybrid-perovskite absorber solar cells. *Energy Environ. Sci.*, **7**, 3690–3698.
46. Leguy, A. M. A., Frost, J. M., McMahon, A. P., Sakai, V. G., Kockelmann, W., Law, C., Li, X. E., Foglia, F., Walsh, A., O'Regan, B. C., Nelson, J., Cabral, J. T., and Barnes, P. R. F. (2015). The dynamics of methylammonium ions in hybrid organic-inorganic perovskite solar cells (vol. 6, 7124, 2015). *Nat. Commun.*, **6**.
47. Yin, W. J., Shi, T. T., and Yan, Y. F. (2014). Unusual defect physics in $CH_3NH_3PbI_3$ perovskite solar cell absorber. *Appl. Phys. Lett.*, **104**.
48. Abate, A., Planells, M., Hollman, D. J., Barthi, V., Chand, S., Snaith, H. J., and Robertson, N. (2015). Hole-transport materials with greatly-differing redox potentials give efficient TiO_2- $[CH_3NH_3][PbX_3]$ perovskite solar cells. *Phys. Chem. Chem. Phys.*, **17**, 2335–2338.
49. Wei, J., Zhao, Y. C., Li, H., Li, G. B., Pan, J. L., Xu, D. S., Zhao, Q., and Yu, D. P. (2014). Hysteresis analysis based on the ferroelectric effect in hybrid perovskite solar cells. *J. Phys. Chem. Lett.*, **5**, 3937–3945.
50. Chen, H. W., Sakai, N., Ikegami, M., and Miyasaka, T. (2015). Emergence of hysteresis and transient ferroelectric response in organo-lead halide perovskite solar cells (vol. 6, 164, 2015). *J. Phys. Chem. Lett.*, **6**, 935–935.
51. Wang, F. F., Meng, D. C., Li, X. N., Zhu, Z., Fu, Z. P., and Lu, Y. L. (2015). Influence of annealing temperature on the crystallization and ferroelectricity of perovskite $CH_3NH_3PbI_3$ film. *Appl. Surf. Sci.*, **357**, 391–396.
52. Chen, B., Shi, J., Zheng, X. J., Zhou, Y., Zhu, K., and Priya, S. (2015). Ferroelectric solar cells based on inorganic-organic hybrid perovskites. *J. Mater. Chem. A*, **3**, 7699–7705.
53. Coll, M., Gomez, A., Mas-Marza, E., Almora, O., Garcia-Belmonte, G., Campoy-Quiles, M., and Bisquert, J. (2015). Polarization switching and light-enhanced piezoelectricity in lead halide perovskites. *J. Phys. Chem. Lett.*, **6**, 1408–1413.
54. Chen, B., Zheng, X., Yang, M., Zhou, Y., Kundu, S., Shi, J., Zhu, K., and Priya, S. (2015). Interface band structure engineering by ferroelectric polarization in perovskite solar cells. *Nano Energy*, **13**, 582–591.

55. Kutes, Y., Ye, L. H., Zhou, Y. Y., Pang, S. P., Huey, B. D., and Padture, N. P. (2014). Direct observation of ferroelectric domains in solution-processed $CH_3NH_3PbI_3$ perovskite thin films. *J. Phys. Chem. Lett.*, **5**, 3335–3339.

56. Wang, Y., Gould, T., Dobson, J. F., Zhang, H. M., Yang, H. G., Yao, X. D., and Zhao, H. J. (2014). Density functional theory analysis of structural and electronic properties of orthorhombic perovskite $CH_3NH_3PbI_3$. *Phys. Chem. Chem. Phys.*, **16**, 1424–1429.

57. Quarti, C., Grancini, G., Mosconi, E., Bruno, P., Ball, J. M., Lee, M. M., Snaith, H. J., Petrozza, A., and De Angelis, F. (2014). The Raman spectrum of the $CH_3NH_3PbI_3$ hybrid perovskite: interplay of theory and experiment. *J. Phys. Chem. Lett.*, **5**, 279–284.

58. Xie, L. Q., Zhang, T. Y., Chen, L., Guo, N. J., Wang, Y., Liu, G. K., Wang, J. R., Zhou, J. Z., Yan, J. W., Zhao, Y. X., Mao, B. W., and Tian, Z. Q. (2016). Organic-inorganic interactions of single crystalline organolead halide perovskites studied by Raman spectroscopy. *Phys. Chem. Chem. Phys.*, **18**, 18112–18118.

59. Ledinsky, M., Loper, P., Niesen, B., Holovsky, J., Moon, S. J., Yum, J. H., De Wolf, S., Fejfar, A., and Ballif, C. (2015). Raman spectroscopy of organic-inorganic halide perovskites. *J. Phys. Chem. Lett.*, **6**, 401–406.

60. Kim, H. S., and Park, N. G. (2014). Parameters affecting I-V hysteresis of $CH_3NH_3PbI_3$ perovskite solar cells: effects of perovskite crystal size and mesoporous TiO_2 layer. *J. Phys. Chem. Lett.*, **5**, 2927–2934.

61. Christians, J. A., Manser, J. S., and Kamat, P. V. (2015). Best practices in perovskite solar cell efficiency measurements. avoiding the error of making bad cells look good. *J. Phys. Chem. Lett.*, **6**, 852–857.

62. Wang, P. Q., Zhao, J. J., Wei, L. Y., Zhu, Q. F., Xie, S. H., Liu, J. X., Meng, X. J., and Li, J. Y. (2017). Photo-induced ferroelectric switching in perovskite $CH_3NH_3PbI_3$ films. *Nanoscale*, **9**, 3806–3817.

63. Jiang, C. S., Yang, M. J., Zhou, Y. Y., To, B., Nanayakkara, S. U., Luther, J. M., Zhou, W. L., Berry, J. J., van de Lagemaat, J., Padture, N. P., Zhu, K., and Al-Jassim, M. M. (2015). Carrier separation and transport in perovskite solar cells studied by nanometre-scale profiling of electrical potential. *Nat. Commun.*, **6**.

64. Dymshits, A., Henning, A., Segev, G., Rosenwaks, Y., and Etgar, L. (2015). The electronic structure of metal oxide/organo metal halide perovskite junctions in perovskite based solar cells. *Sci. Rep.*, **5**.

65. Sewvandi, G. A., Hu, D. W., Chen, C. D., Ma, H., Kusunose, T., Tanaka, Y., Nakanishi, S., and Feng, Q. (2016). Antiferroelectric-to-ferroelectric

switching in $CH_3NH_3PbI_3$ perovskite and its potential role in effective charge separation in perovskite solar cells. *Phys. Rev. Appl.*, **6**.

66. Rakita, Y., Bar-Elli, O., Meirzadeh, E., Kaslasi, H., Peleg, Y., Hodes, G., Lubomirsky, I., Oron, D., Ehre, D., and Cahen, D. (2017). Tetragonal $CH_3NH_3PbI_3$ is ferroelectric. *Proc. Natl. Acad. Sci. U.S.A.*, **114**, E5504–E5512.

67. Gonzalez-Carrero, S., Frances-Soriano, L., Gonzalez-Bejar, M., Agouram, S., Galian, R. E., and Perez-Prieto, J. (2016). The luminescence of $CH_3NH_3PbBr_3$ perovskite nanoparticles crests the summit and their photostability under wet conditions is enhanced. *Small*, **12**, 5245–5250.

Chapter 5

Tandem Structure

Hiroyuki Kanda, Naoyuki Shibayama, and Seigo Ito
Department of Materials and Synchrotron Radiation Engineering, Graduate School of Engineering, University of Hyogo. 2167 Shosha, Himeji, Hyogo, 671-2280, Japan
itou@eng.u-hyogo.ac.jp

5.1 Introduction

Photovoltaic (PV) energy has become quite cost effective, with 0.05–0.38 A$/KWh [1]. However, steady effort is important for continuous cost reduction. In any case, PV electricity can't become cheaper easily even if the solar cell became cheaper, because there are other costs, like the cost of module processing, flame material, stage, and cable and the setting fee. A direct method to lower the total cost per generation electricity is to raise the conversion efficiency of the system. In other words, since PV cell cost is roughly 1/3 of the total cost, when we reduce the cell price by half, the cost per generation electricity goes down by approximately 1/6. However, if we can double the conversion efficiency, the system price per generation electricity can be reduced to half directly.

Multifunctional Organic–Inorganic Halide Perovskite: Applications in Solar Cells, Light-Emitting Diodes, and Resistive Memory
Edited by Nam-Gyu Park and Hiroshi Segawa

ISBN 978-981-4800-52-5 (Hardcover), 978-1-003-27593-0 (eBook)
www.jennystanford.com

At present, it has been reported that the best conversion efficiencies of perovskite, silicon heterojunction, and copper indium gallium selenide (CIGS) solar cells as single-junction solar cells are 22.1%, 26.6%, and 22.6%, respectively, under 1 sun irradiation (the global AM 1.5 spectrum [1000 Wm^{-2}] at 25°C, IEC 60904-3: 2008, ASTM G-173-03 global) [2]. The conversion efficiency of single-junction solar cells is limited to around 30% due to the wide distribution of the solar irradiation energy and internal thermal loss in the semiconductors (i.e., Shockley–Queisser limit [3]).

The photoenergy conversion efficiency can be improved by dividing the solar irradiation energy to the semiconductors with several bandgaps, resulting in minimization of the internal thermal loss. To improve the conversion efficiency, a tandem structure can be quite significant. A tandem structure has the wide bandgap solar cell at the front side, against the sun, (the top cell) and the narrow bandgap one at the opposite side (the bottom cell). The top and bottom cells absorb high- (blue) and low-energy (red) lights, respectively (Fig. 5.1). If the number of cells increases to infinite, the conversion efficiency can be improved to 69.9%, theoretically [3]. Now, by the Shockley–Queisser limit, theoretically, although the maximal conversion efficiency of single-junction silicon solar cells is 33.5% for the AM 1.5 G spectrum at 25°C, the conversion efficiencies of the tandem cells using silicon and perovskite are 45.1% and 45.3% for the two- and four-terminal structures, respectively (Fig. 5.2) [3]. The possible conversion efficiency of the tandem solar cell was estimated to be over 31% from a practical point of view [4].

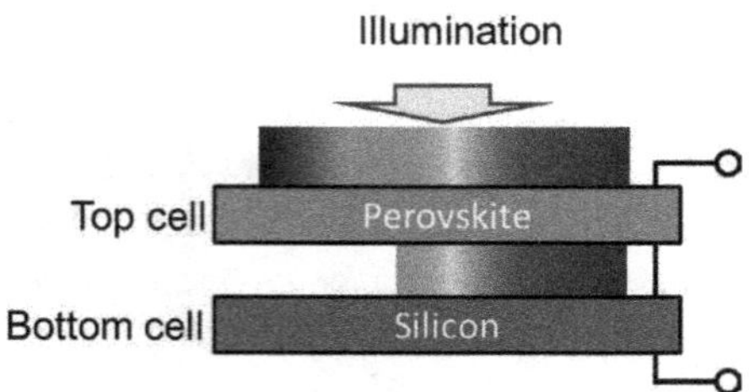

Figure 5.1 Schematic structure of a tandem solar cell using two junctions (perovskite and silicon).

Therefore, significant improvement is expected in the photoenergy conversion efficiency of the tandem solar cells using perovskite and silicon junctions. Figure 5.3 shows the efficiency progress chart of tandem solar cells <perovskite/silicon>, <perovskite/perovskite>, and <perovskite/CIGS>. It can be noticed that a tandem structure has an advantage in the photoenergy conversion efficiency against single-junction perovskite solar cells. In this chapter, the progress in the structure of each tandem devise and the technological points of fabrication have been described.

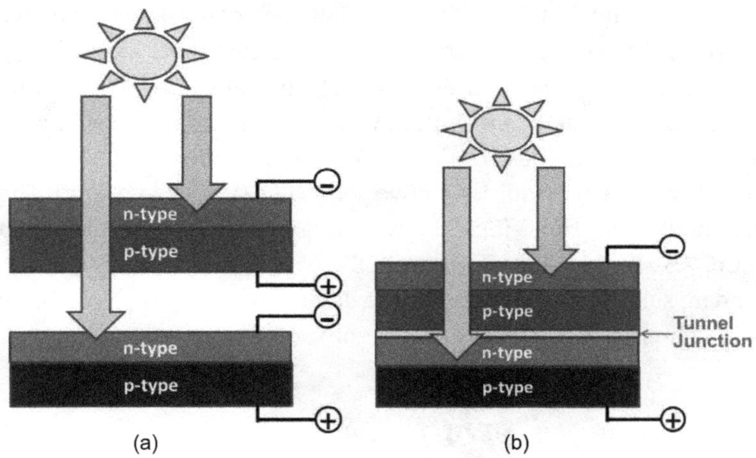

Figure 5.2 Schematic structure of a (a) four- and (b) two-terminal tandem solar cell using two junctions (perovskite and silicon).

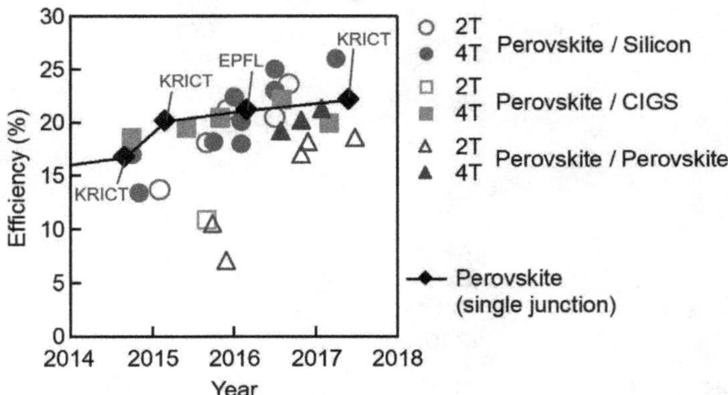

Figure 5.3 The efficiency progress chart of tandem solar cells <perovskite/silicon>, <perovskite/perovskite>, and <perovskite/CIGS> compared with the best efficiency of perovskite solar cells.

5.2 Beam Splitting System

Figure 5.4 shows the measurement scheme of an optical tandem system using a beam splitter, which can divide the light of the solar simulator (AM 1.5) into two parts, one with long wavelengths and one with short wavelengths. Table 5.1 shows the activities of tandem solar cells using a beam splitter using various combinations of top perovskite and bottom solar cells (silicon, perovskites, and dye sensitized) [5–7]. The results of the open-circuit photovoltage (V_{oc}), short-circuit photocurrent density (J_{sc}), and fill factor (FF) can't be obtained using the beam splitter, which is the same condition with the four-terminal ones. The power conversion efficiency (PCE) can be obtained by the addition of two solar cells (top and bottom). The maximal conversion efficiency was obtained, using perovskite top (PCE: 7.5%) and silicon bottom (PCE: 20.5%) cells, to be 28% as the tandem solar cell [5], which can be a demonstration of the possibility of a high-efficiency conversion system.

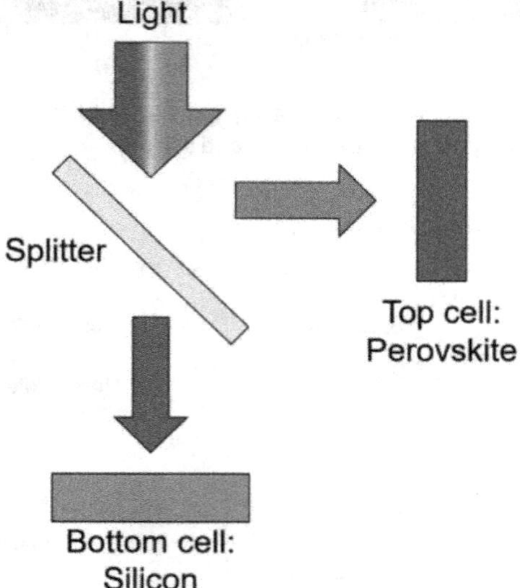

Figure 5.4 The measurement scheme of a tandem (perovskite + silicon) solar cell using a beam splitter.

Table 5.1 Reported tandem solar cell with a perovskite solar cell using the beam splitting system

Structure (top and bottom cells)	J_{SC} (mA cm^{-2})	V_{OC} (V)	FF	Eff. (%)	Ref.
FTO/TiO_2/mpTiO_2-$MAPbI_3$/Spiro/Au	10.6	0.987	0.715	7.5	[5]
MgF_2/TCO/a-Si:H(p^+)/a-Si:H(i)/c-Si(n)/a-Si:H(i)/a-Si:H(n^+)/TCO/Ag	34.9	0.728	0.809	20.5	
Tandem (perovskite/silicon)	–	–	–	28.0	
FTO/TiO_2/mpTiO_2-$MAPbBr_3$/Spiro/Au	7.2	1.265	0.720	6.5	[6]
PERL silicon solar cell	31.6	0.673	0.800	16.9	
Tandem (perovskite/silicon)	-	–	–	23.4	
FTO/TiO_2/mpTiO_2-$MAPbBr_3$/Spiro/Au	7.2	1.265	0.720	6.5	[6]
FTO/TiO_2/$MAPbI_3$/Spiro/Au	12.1	0.826	0.680	6.9	
Tandem (perovskite/perovskite)	-	–	–	13.4	
FTO/TiO_2/mpTiO_2-$(FAPbI_3)_{0.85}(MABr_3)_{0.15}$/PTAA/Au	20.6	1.120	0.795	18.3	[7]
Dye-sensitized solar cell	8.98	0.507	0.694	3.2	
Tandem (perovskite/dye sensitized)	-	–	–	21.5	

5.3 Perovskite/Silicon

5.3.1 Four-Terminal Perovskite/Silicon

Table 5.2 summarizes the research activity of four-terminal tandem solar cells (Fig. 5.2a [4, 8–18]. For the first time, Löper et al. [4] published the four-terminal tandem solar cells, which were composed of n-i-p-type perovskite solar cells and silicon heterojunction solar cells (Fig. 5.5). The structure of silicon heterojunction solar cells were <indium-tin oxide (ITO)/amorphous silicon (a-Si) p-type layer (a-Si:H(p))/a-Si i-type layer (a-Si:H(i))/crystal n-type silicon wafer/a-Si i-type layer (a-Si:H(i))/a-Si n-type layer (a-Si:H(n))/ITO/Ag>.

In the perovskite solar cells, MoO*x* layers were composed in order to reduce the ITO sputtering damage on the organic hole conductor (Spiro-OMeTAD), which can improve the photocurrent density (J_{sc}) from 4 mA cm^{-2} to 4.5 mA cm^{-2}, resulting in the conversion efficiencies of 6.2%, 7.2%, and 13.4% of the perovskite, light-filtered silicon, and tandem solar cells, respectively.

Table 5.2 Reported four-terminal perovskite/silicon tandem solar cell

Structure (top and bottom cells)	J_{SC} **(mA cm^{-2})**	V_{OC} **(V)**	**FF**	**Eff. (%)**	**Ref.**
LiF/FTO/TiO_2/mpTiO_2-$MAPbI_3$/Spiro/Ag NWs/LiF	17.5	1.025	0.710	12.7	[8]
SiN*x*/emitter/mc-Si/Al	11.1	0.547	0.704	4.3	
Tandem	–	–	–	17.0	
FTO/TiO_2/mpTiO_2-$MAPbI_3$/Spiro/MoO*x*/ITO	14.5	0.821	0.519	6.2	[4]
ITO/a-Si:H(p^+)/a-Si:H(i)/c-Si(n)/a-Si:H(i)/a-Si:H(n^+)/ITO/Ag	13.7	0.689	0.767	7.2	
Tandem	–	–	–	13.4	
MgF_2/ITO/TiO_2/mpTiO_2-$MAPbI_3$/Spiro/MoO*x*/ITO/MgF_2	18.8	0.950	0.690	12.2	[9]
PERL silicon solar cell	16.9	0.640	0.730	7.9	
Tandem	–	–	–	20.1	
FTO/TiO_2/mpTiO_2-$MAPbI_3$/Spiro/MoO*x*/IZO	17.5	0.870	0.680	10.4	[10]
a-Si:H/c-Si(n)	14.6	0.690	0.776	7.8	
Tandem	–	–	–	18.2	
ITO/IO:H/MoO*x*/Spiro/$MAPbI_3$/PCBM/PEIE/IZO	20.1	1.070	0.755	16.2	[11]
ITO/a-Si:H(p^+)/a-Si:H(i)/c-Si(n)/a-Si:H(i)/a-Si:H(n^+)/ITO/Ag	16.0	0.693	0.795	8.8	
Tandem	–	–	–	25.0	

Structure (top and bottom cells)	J_{SC} (mA cm^{-2})	V_{OC} (V)	FF	Eff. (%)	Ref.
MgF_2/ITO/PEDOT:PSS/ $MAPbI_3$/PCBM/AZO/ITO/ MgF_2	16.5	0.952	0.774	12.3	[12]
c-Si cell	13.3	0.562	0.762	5.7	
Tandem	–	–	–	18.0	
ITO/PTAA/MAPbI3/PCBM/ C60/BCP/Cu/Au/BCP	20.6	1.080	0.741	16.5	[13]
IZO/a-Si:H(p^+)/a-Si:H(i)/c-Si(n)/a-Si:H(i)/a-Si:H(n^+)/ ITO/MgF_2/Ag	12.3	0.679	0.779	6.5	
Tandem	–	–	–	23.0	
FTO/SnO_2-PCBM/ $FA_{0.83}Cs_{0.17}PbI_{1.8}Br_{1.2}$/Spiro/ ITO	19.9	1.100	0.707	15.1	[14]
ITO/a-Si:H(p^+)/a-Si:H(i)/c-Si(n)/a-Si:H(i)/a-Si:H(n^+)/ AZO/Ag	13.9	0.690	0.764	7.3	
Tandem	–	–	–	22.4	
ITO/TiO_2/Rb-$FA_{0.75}(MA_{0.6}Cs_{0.4})_{0.25}PbI_2Br$/ PTAA/$MoO_x$/ITO/$MgF_2$	19.4	1.130	0.700	16.0	[15]
SiO_x/SiN_x/c-Si(n)/emitter (p)/ SiO_2/Si_3N_4/Al	18.8	0.690	0.800	10.4	
Tandem	–	–	–	26.4	
FTO/TiO_2/$MAPbI_3$/Spiro/ graphene	12.6	0.900	0.550	6.2	[16]
a-Si:H/c-Si	14.0	0.670	0.738	7.0	
Tandem	–	–	–	13.2	
FTO/TiO_2/mpTiO_2/$MAPbI_3$/ Spiro/thin-Au/ITO	16.8	1.028	0.581	10.0	[17]
ITO/emitter(n)/p-Si/Al	12.3	0.520	0.689	4.4	
Tandem	–	–	–	14.4	
FTO/TiO_2/mpTiO_2/$MAPbI_3$/ Spiro/MoO_X/IZO	22.5	0.960	0.733	15.9	[18]
ITO/emitter(n)/p-Si/Al	16.1	0.464	0.528	3.9	
Tandem	–	–	–	19.8	

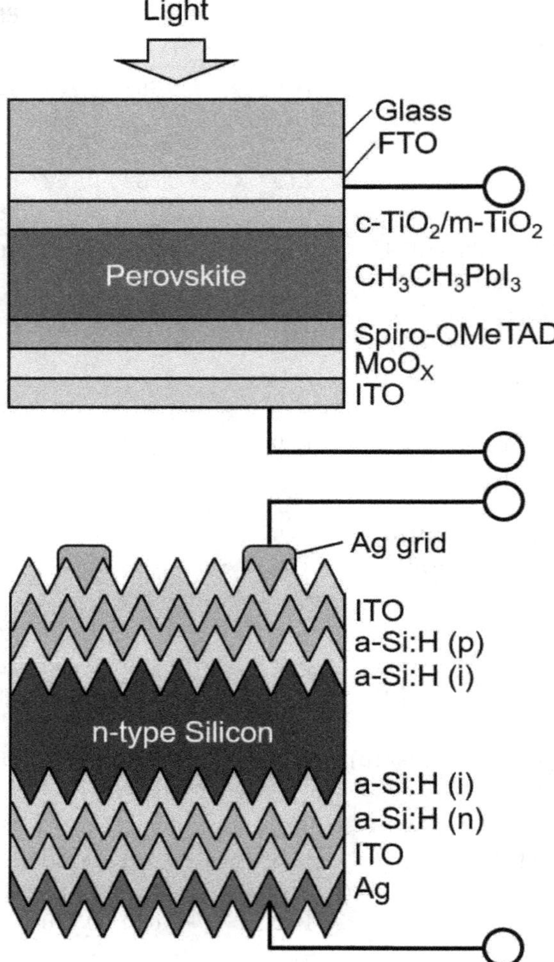

Figure 5.5 Four-terminal structured perovskite/silicon tandem solar cells (modified from Ref. [4]).

Duong et al. reported the top conversion efficiency of perovskite/silicon tandem solar cells by tuning the bandgap of perovskite with the mixing variation of Ru^+, Cs^+, $CH_3NH_3{}^+$, $(NH_2)_2CH^+$, Pb^{2+}, I^-, and Br^-, which can improve the J_{SC} of bottom silicon cells [15]. The resulting absorption edge was 720 nm. The addition of Ru^+ cation into the perovskite crystal improved the crystallinity and hysteresis (from 0.2 to 0). The conversion efficiencies were 16%, 10.4%, and 26.4% of the perovskite, light-filtered silicon, and tandem solar cells, respectively.

5.3.2 Two-Terminal Perovskite/Silicon

Fabrication of two-terminal tandem solar cells using a perovskite layer is quite difficult technically because of the damage in the substrate layers during the coating of the additional upper layers. The coating method for each layer in the two-terminal tandem solar cells is also summarized in the tables below.

Table 5.3 summarizes two-terminal perovskite/silicon solar cells [17–23]. For the first time, Mailoa et al. [19] published using n-i-p-type perovskite solar cells <$TiO_2/CH_3NH_3PbI_3$/Spiro> and n-type wafer silicon solar cells (Fig. 5.6a). Specially, this structure was unique due to the tunneling junction between n^{++} a-Si and p^{++} a-Si, fabricated by plasma-enhanced chemical vapor deposition (Fig. 5.6b). And then, for the transparent conducting layer, Ag nanowire was coated on top by low-temperature deposition to eliminate thermal damage on perovskite solar cells. The conversion efficiency was 13.7% as the 2-terminal tandem solar cells.

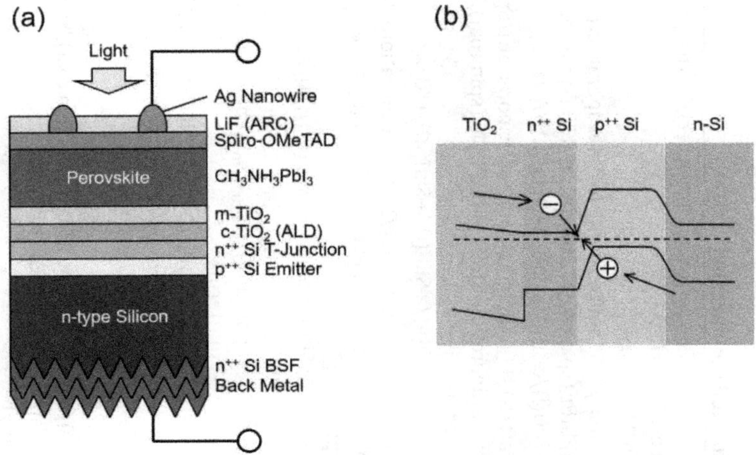

Figure 5.6 Structure of a two-terminal perovskite/silicon tandem solar cell (a) and the energy band diagram of the tunnel junction (b) (modified from Ref. [19]).

Table 5.3 Reported two-terminal perovskite/silicon tandem solar cell

Structure (top and bottom cells)	Assembling method	J_{SC} (mA cm^{-2})	V_{OC} (V)	FF	Eff. (%)	Ref.
LiF (evaporation)/Ag NWs (mechanical transfer)/Spiro (spin coat)/ mpTiO$_2$-MAPbI$_3$ (spin coat)/TiO$_2$ (ALD) a-Si:H/a-Si/p^{++} Si/n-Si/n^{++} Si/metal	Monolithic	11.5	1.580	0.750	13.7	[19]
LiF (evaporation)/ITO (sputtering)/MoO$_3$ (evaporation)/Spiro (spin coat)/FAMAPbIBr (spin coat)/SnO$_2$ (ALD)/ITO (sputtering) a-Si:H(p$^+$)/a-Si:H(i)/c-Si(n)/a-Si:H(i)/a-Si:H(n$^+$)/AZO/Ag	Monolithic	12.9	1.780	0.790	18.1	[20]
ITO (spin coat)/IO:H (sputtering)/MoOx (evaporation)/Spiro (spin coat)/MAPbI$_3$ (evaporation/spin coat)/PCBM (spin coat)/PEIE (spin coat)/IZO (sputtering) a-Si:H(p$^+$)/a-Si:H(i)/c-Si(n)/a-Si:H(i)/a-Si:H(n$^+$)/ITO/Ag	Monolithic	15.9	1.690	0.776	21.2	[21]
ITO (sputtering)/IO:H (sputtering)/MoOx (evaporation)/Spiro (spin coat)/MAPbI$_3$ (evaporation/spin coat)/PCBM (spin coat)/PEIE (spin coat)/IZO (sputtering) a-Si:H(p$^+$)/a-Si:H(i)/c-Si(n)/a-Si:H(i)/a-Si:H(n$^+$)/ITO/Ag	Monolithic	16.4	1.717	0.731	20.5	[22]
LiF (evaporation)/ITO (sputtering)/SnO$_2$-ZTO (CVD)/PCBM (evaporation)/LiF (evaporation)/FA$_{0.83}$Cs$_{0.17}$PbI$_{2.49}$Br$_{0.51}$ (spin coat)/ NiO (spin coat)/ITO (sputtering) a-Si:H(n$^+$)/a-Si:H(i)/c-Si(n)/a-Si:H(i)/a-Si:H(p$^+$)/ITO/Si NP/Ag	Monolithic	18.1	1.650	0.790	23.6	[23]
FTO/TiO$_2$ (SPD)/mpTiO$_2$ (spin coat)/MAPbI$_3$ (spin coat)/Spiro (spin coat)/thin-Au (evaporation)/ITO (sputtering) ITO/emitter(n)/p-Si/Al	Mechanical	12.3	1.564	0.713	13.7	[17]
FTO/TiO$_2$ (SPD)/mpTiO$_2$ (spin coat)/MAPbI$_3$ (spin coat)/Spiro (spin coat)/MoO$_X$ (evaporation)/IZO (sputtering) ITO/emitter(n)/p-Si/Al	Monolithic	16.3	1.420	0.666	15.5	[18]

Werner et al. fabricated a tandem device with silicon heterojunction solar cells [21] (Fig. 5.7). The silicon heterojunction can improve the V_{OC} due to the surface passivation effect by an a-Si layer on a crystal silicon wafer. However, since the a-Si can deteriorate on thermal treatment over 200°C, the processing temperature of a tandem device on silicon heterojunction solar cells should be below 200°C. Therefore, they used an n-type organic electron transportation material (PCBM) in the perovskite solar cells with low-temperature processing (<150°C), resulting in the 21.2% conversion efficiency as the 2-terminal perovskite solar cell.

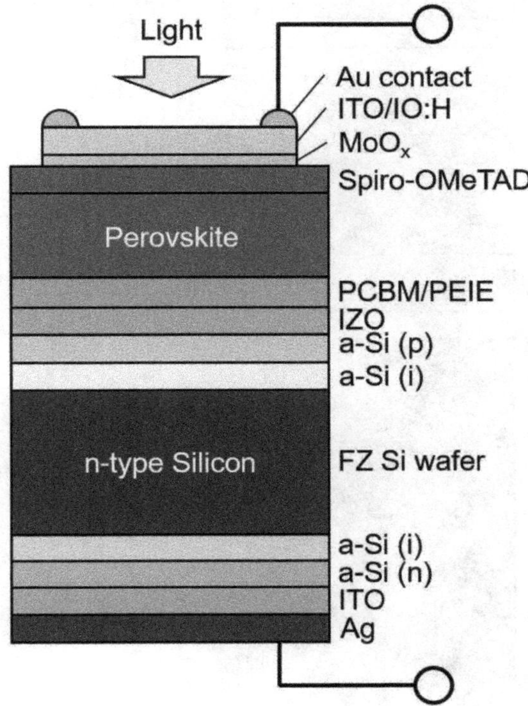

Figure 5.7 Structure of two-terminal perovskite/silicon tandem solar cells using silicon heterojunction solar cells (modified from Ref. [21]).

Since crystal silicon has a low absorption coefficient, it is important to enlarge the optical path length in the device. To improve the photocurrent density by enhancement of the optical path length,

Bush et al. utilized silicon nanoparticles on the backside of bottom silicon solar cells for the near-infrared (NIR)-light reflection layer (Fig. 5.8) [23]. The deposition of silicon nanoparticles enhanced the light reflection of 99% at the backside in silicon solar cells, resulting in the J_{SC} improvement of 1.5 mA cm^{-2} and a high conversion efficiency of 23.6%.

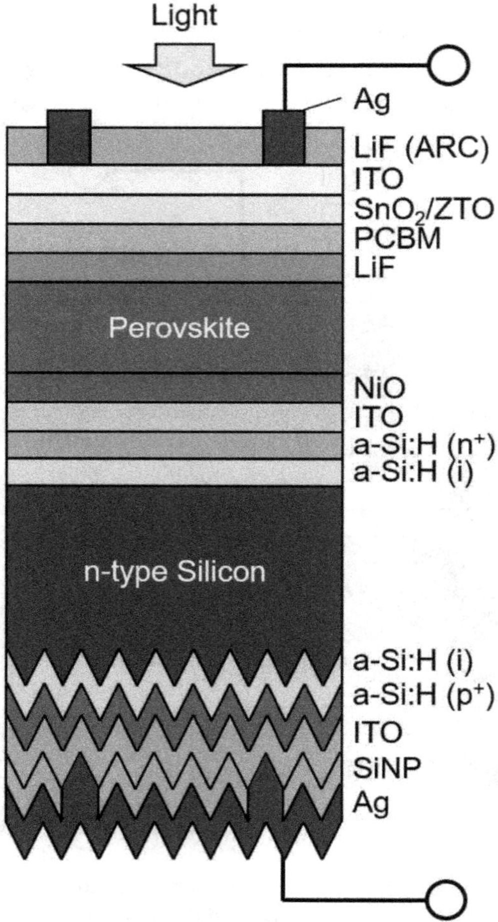

Figure 5.8 Structure of two-terminal perovskite/silicon tandem solar cells using silicon nanoparticles (Si NPs) on the back for the NIR-light reflection layer (modified from Ref. [23]).

5.4 Perovskite/CIGS

$Cu(In_xGa_{1-x})(Se_yS_{1-y})_2$ solar cells (CIGS) can be bottom cells in tandem cells using top perovskite solar cells. The PV results of four- and two-terminal tandem solar cells are summarized in Tables 5.4 and 5.5, respectively, [8, 24–30]. Specially, CIGS can be fabricated on a flexible substrate by roll-to-roll processing. Moreover, the bandgap of CIGS can be tuned from 1.10 eV to 1.24 eV by changing the elemental ratio, which is useful for the optimization of the light absorption edge.

Table 5.4 Reported four-terminal perovskite/CIGS tandem solar cell

Structure (top and bottom cells)	J_{SC} (mA cm^{-2})	V_{OC} (V)	FF	Eff. (%)	Ref.
LiF/FTO/TiO_2/mpTiO_2-$MAPbI_3$/Spiro/Ag NWs/LiF	17.5	1.025	0.710	12.7	[8]
CIGS	10.9	0.682	0.788	5.9	
Tandem	–	–	–	18.6	
FTO/TiO_2/$MAPbI_3$/Spiro/ Ag NWs	17.2	0.892	0.540	8.3	[24]
Ag NWs/AZO/i-ZnO/CdS/ CIGS/Mo	11.2	0.500	0.446	2.5	
Tandem	–	–	–	10.8	
FTO/TiO_2/mpTiO_2-$MAPbI_3$/ Spiro/MoO_3/ZnO:Al	16.7	1.034	0.703	12.1	[25]
ZnO/ZnO:Al/CdS/CIGS/Mo	14.4	0.661	0.774	7.4	
Tandem	–	–	–	19.5	
MoOx/Au/Ag/MoOx/Spiro/ $MAPbI_3$/TiO_2/ITO	14.6	1.050	0.751	11.5	[26]
ITO/CdS/CIGS/Mo	10.2	0.560	0.696	4.0	
Tandem	–	–	–	15.5	
FTO/ZnO/PCBM/$MAPbI_3$/ Spiro/MoO_3/In_2O_3:H	17.4	1.104	0.736	14.1	[27]
i-ZnO/ZnO:Al/CdS/CIGS/Mo	12.7	0.667	0.749	6.3	
Tandem	–	–	–	20.4	

(Continued)

Table 5.4 (*Continued*)

Structure (top and bottom cells)	J_{SC} (mA cm^{-2})	V_{OC} (V)	FF	Eff. (%)	Ref.
ZnO:Al/ZnO/PCBM/$MAPbI_3$/PTAA/In_2O_3:H	19.1	1.116	0.754	16.1	[28]
i-ZnO/ZnO:Al/CdS/CIGS/Mo	12.1	0.669	0.736	6.0	
Tandem	–	–	–	22.1	
FTO/TiO_2/mpTiO_2-$MAPbI_3$/Spiro/Ag/ITO	20.1	0.975	0.781	15.3	[29]
ITO/CdS/CIGS/Mo	15.2	0.470	0.646	4.6	
Tandem	–	–	–	19.9	

5.4.1 Four-Terminal Perovskite/CIGS

Bailie et al. have published four-terminal perovskite/silicon tandem solar cells for the first time (Fig. 5.9) [8]. To ensure that the NIR light passes to the bottom CIGS, Ag nanowire has been utilized as the back electrode on perovskite solar cells, which has transparency and high conductivity (sheet resistivity, 12.4 Ω sp^{-1}; transparency: 87%–90%). The resulting conversion efficiencies of transparent perovskite, filtered CIGS, and tandem cell were 12.7%, 5.9%, and 18.6%, respectively.

Fu et al. utilized PTAA and PCBM-ZnO for hole-transporting material (HTM) and electron-transporting material (ETM) on the perovskite layer, respectively (Fig. 5.10), which has stabilized conversion efficiency against temperature variation (–0.18% °C^{-1} from 25°C to 65°C) [28]. The resulting conversion efficiencies of transparent perovskite, filtered CIGS, and four-terminal tandem cell were 16.1%, 6%, and 22.1%, respectively.

5.4.2 Two-Terminal Perovskite/CIGS

Since the CIGS can deteriorate due to thermal stress over 200°C, the two-terminal tandem device should be fabricated below 200°C. Hence, for the first time, Todorov et al. fabricated two-terminal perovskite/CIGS tandem cells using PCBM and PEDOT:PSS as the ETM and HTM, respectively, in the perovskite solar cells for the low-temperature procedure (Table 5.5) [30].

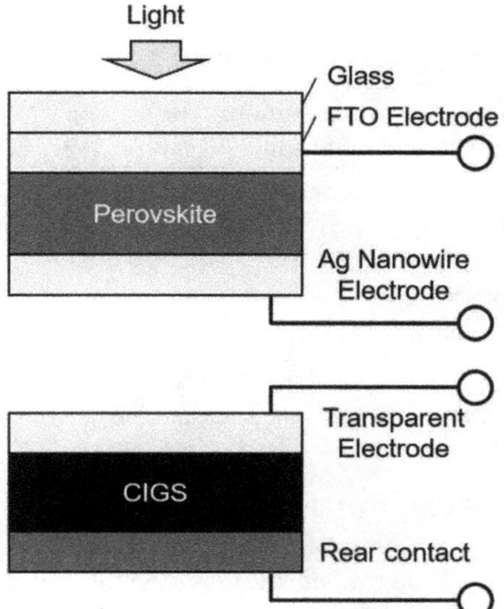

Figure 5.9 Four-terminal perovskite/CIGS tandem solar cells using Ag nanowire on the transparent perovskite back electrodes (modified from Ref. [8]).

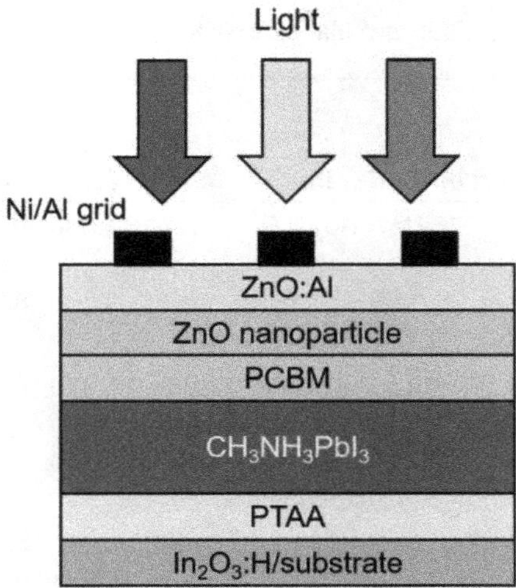

Figure 5.10 Transparent perovskite solar cells using PTAA as HTM and PCBM/ZnO nanoparticle as ETM for the utilization of a four-terminal tandem device on CIGS solar cells (modified from Ref. [28]).

Table 5.5 Reported two-terminal perovskite/CIGS tandem solar cell

Structure (top and bottom cells)	Assembling method	J_{SC} (mA cm^{-2})	V_{OC} (V)	FF	Eff. (%)	Ref.
Ca (evaporation)/ BCP (evaporation)/ PCBM (sputtering)/ $MAPbI_xBr_{3-x}$ (spin coat)/ PEDOT:PSS (spin coat)/ ITO (sputtering) $CdS/CIGS/Mo/Si_3N_4$	Monolithic	12.7	1.45	0.566	10.9	[30]

5.5 Perovskite/Perovskite

Since the bandgap of the perovskite crystal can be tuned from 1.2 eV to 2.3 eV by Br and Sn doping, tandem cells using two perovskite layers with low and high bandgaps can be fabricated (Tables 5.6 and 5.7 [31–37]). The resulting high-bandgap perovskite/low-bandgap perovskite tandem cells can be fabricated by the roll-to-roll process.

Table 5.6 Reported four-terminal perovskite/perovskite tandem solar cell

Structure (top and bottom cells)	J_{SC} (mA cm^{-2})	V_{OC} (V)	FF	Eff. (%)	Ref.
$ITO/NiO_x/MAPbI_3/PCBM/C60/ITO$	16.7	1.080	0.750	13.5	[31]
ITO/PEDOT:PSS/ $FA_{0.5}MA_{0.5}Pb_{0.}75Sn_{2.1}I_3/PCBM/C60/Ag$	9.1	0.760	0.800	5.6	
Tandem	–	–	–	19.1	
$ITO/NiO/FA_{0.83}Cs_{0.17}Pb(I_{0.83}Br_{0.17})_3/$ SnO_2-PCBM/ITO	20.3	0.970	0.790	15.7	[32]
ITO/PEDOT:PSS/ $FA_{0.75}Cs_{0.25}Pb_{0.5}Sn_{0.5}I_3$/C60-BCP/Ag	7.9	0.740	0.730	4.4	
Tandem	–	–	–	20.1	
$MoO_x/Au/MoO_x$/Spiro/ $FA_{0.3}MA_{0.7}PbI_3/C60/SnO_2/FTO$	20.1	1.141	0.800	18.3	[33]
ITO/PEDOT:PSS/ $FA_{0.6}MA_{0.7}Pb_{0.4}Sn_{0.6}I_3$/C60-BCP/Ag	4.8	0.808	0.744	2.9	
Tandem	–	–	–	21.2	

Table 5.7 Reported two-terminal perovskite/perovskite tandem solar cell

Structure (top and bottom cells)	Assembling method	J_{SC} (mA cm^{-2})	V_{OC} (V)	FF	Eff. (%)	Ref.
FTO/TiO_2 (spray pyrolysis deposition)/$MAPbBr_3$ (spin coat)/P3HT (spin coat) PCBM (spin coat)/$MAPbI_3$ (spin coat)/PEDOT:PSS (spin coat)/ITO	Mechanical	8.4	1.95	0.66	10.8	[34]
PEDOT:PSS (spin coat)/Spiro (spin coat)/$MAPbI_3$ (spin coat)/PCBM (spin coat)/PEI (spin coat) PEDOT:PSS (spin coat)/Spiro (spin coat)/mpTiO_2-$MAPbI_3$ (spin coat)/TiO_2 (spin coat)/FTO	Monolithic	6.6	1.890	0.560	7.0	[35]
ITO/IPH (spin coat)/$FA_{0.85}Cs_{0.15}PbI_{0.9}Br_{2.1}$ (spin coat)/TaTm (evaporation)/TaTm:F6-TCNNQ (evaporation) C60:Phlm (evaporation)/C60 (evaporation)/$MAPbI_3$ (evaporation)/TaTm (evaporation)/TaTm:F6-TCNNQ (evaporation)/Au (evaporation)	Monolithic	9.8	2.294	0.803	18.1	[36]
ITO/NiO (spin coat)/$FA_{0.83}Cs_{0.17}PbI_{1.5}Br_{1.5}$ (spin coat)/PCBM (evaporation)/SnO_2 (ALD)/ITO (sputtering) PEDOT:PSS (spin coat)/$FA_{0.75}Cs_{0.25}Pb_{0.5}Sn_{0.5}I_3$ (spin coat)/C60-BCP (spin coat)/Ag (evaporation)	Monolithic	14.5	1.660	0.700	17.0	[32]
ITO/NiO_x (spin coat)/$MA_{0.9}Cs_{0.1}PbI_{1.8}Br_{1.2}$ (spin coat)/C60 (evaporation)/Bis-C60 (spin coat)/ITO (sputtering) PEDOT:PSS (spin coat)/$MAPb_{0.5}Sn_{0.5}I_3$ (spin coat)/IC60BA (spin coat)/Bis-C60 (spin coat)/Ag (evaporation)	Monolithic	12.7	1.980	0.730	18.5	[37]

5.5.1 Four-Terminal Perovskite/Perovskite

Yang et al. fabricated four-terminal high-bandgap perovskite/low-bandgap perovskite solar cells for the first time (Fig. 5.11) [31]. Sn was doped in the perovskite layer in order to narrow the bandgap, and then, the absorption edge was shifted from 800 nm to 950 nm. The resulting conversion efficiencies of the transparent high-bandgap perovskite, filtered low-bandgap perovskite, and four-terminal tandem cell, were 13.5%, 5.6%, and 19.1%, respectively.

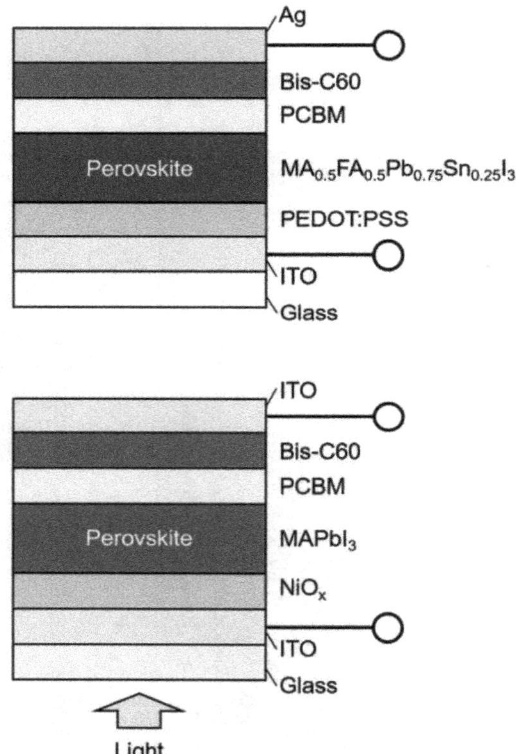

Figure 5.11 Four-terminal high-bandgap perovskite/low-bandgap perovskite solar cells. The bandgap narrowing from 800 nm to 950 nm was performed by Sn doping in the organo-metal-halide perovskite.

5.5.2 Two-Terminal Perovskite/Perovskite

Eperon et al. fabricated two-terminal perovskite/perovskite solar cells using an ITO junction layer (Fig. 5.12) [32]. Since perovskite layers can be dissolved in polar solvents, a spin-coating deposition of another layer on perovskite can make damage in the perovskite layer below [38]. Therefore, deposition of inorganic materials (SnO_2 and ITO) on the perovskite layer is important to reduce the damage by the deposition of another perovskite layer as the tandem device. At first, SnO_2 was deposited by ALD on PCBM, and then, ITO was sputtered on the SnO_2, which can be a good barrier layer against polar solvents to deposit perovskite. The resulting conversion efficiency of 2-terminal solar cells was 17.0%.

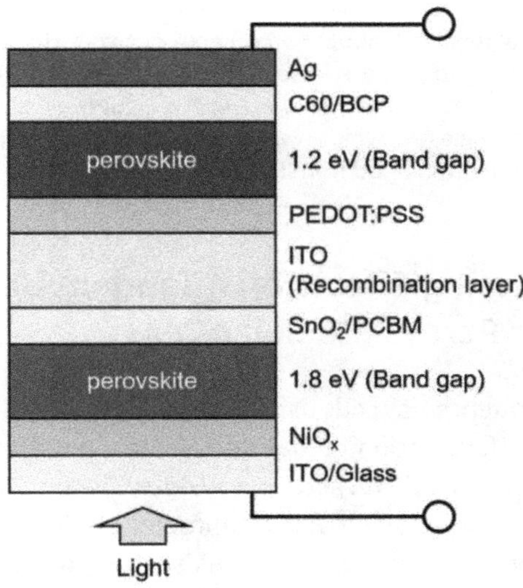

Figure 5.12 Structure of two-terminal high-bandgap perovskite/low-bandgap perovskite solar cells using an ITO layer as the tunnel junction between the two solar cells (modified from Ref. [32]).

To eliminate the damage on perovskite due to spin coating, Heo et al. fabricated two-terminal perovskite/perovskite solar cells by face-to-face gluing between two perovskite solar cells (Fig. 5.13) [34], resulting in 10.8% conversion efficiency.

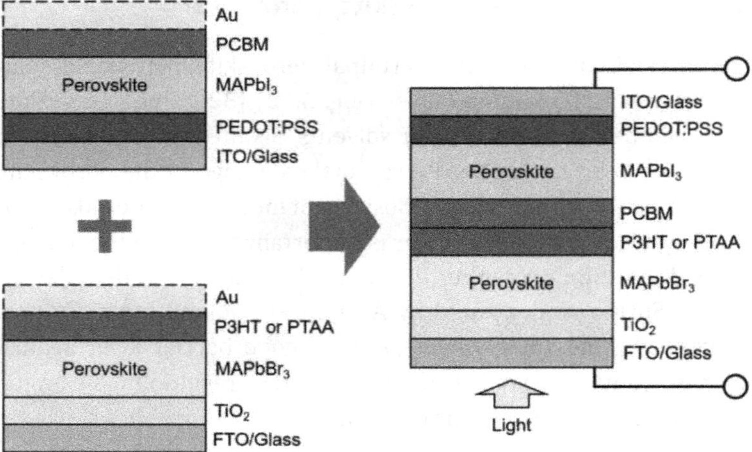

Figure 5.13 Structure and fabrication scheme of two-terminal high-bandgap perovskite/low-bandgap perovskite solar cells by face-to-face gluing between two perovskite solar cells. The Au layers in the figures on the left were the electrodes for the single cell, which were eliminated in the figure on the right as the tandem structure (modified from Ref. [34]).

5.6 Processing Effects on a Tandem Device Using Perovskite Solar Cells

To fabricate tandem solar cells using a perovskite layer, it is important to establish the fabrication guideline. Our group has fabricated and optimized mechanical-stacked perovskite/silicon tandem solar cells and understood some parts of optoelectronics at the interface between perovskite and silicon. In this section, our understanding about the junction layers has been described.

5.6.1 ITO Sputtering Damage in a Hole Transport Material [39]

It is necessary for the top perovskite solar cells to pass the long-wavelength light to the bottom cells in a tandem structure. Therefore, the back-contact electrodes at the top of the device need to be transparent. In reports of tandem solar cells, $In_xSn_yO_3$ (ITO) has been utilized frequently.

However, the ITO layer can be deposited by sputtering, basically, which can inflict sputtering damages in an organic hole conducting material, and then, the resulting S-shaped photocurrent-voltage (*I*-*V*) (Fig. 5.14). To analyze the sputtering damage of ITO on the HTM, the photocurrent-voltage curves of the device on changing the ITO sputtering time have been measured and fitted (as Fig. 5.15) using equivalent circuits with three diodes (containing a reverse diode) (Fig. 5.16). It was confirmed that the twist of *I*-*V* curves to S shape was enhanced by the ITO sputtering time and that the analysis using the equivalent circuit cleared the following two points by ITO sputtering on the HTM layer:

(1) Structural damage in HTM layer
(2) Physical damage making the Schottky barrier between ITO and HTM

About (1), the shrinkage of HTM layer by ITO sputtering was confirmed in the cross-sectional SEM image of perovskite solar cells (Fig. 5.16a). About (2), a numerical analysis showed that the Shottky barrier increased with increase of the ITO sputtering time. Therefore, it was confirmed that the elimination of sputtering damage is an important issue for the improvement of photovoltaic effect on the tandem cells using HTM [39].

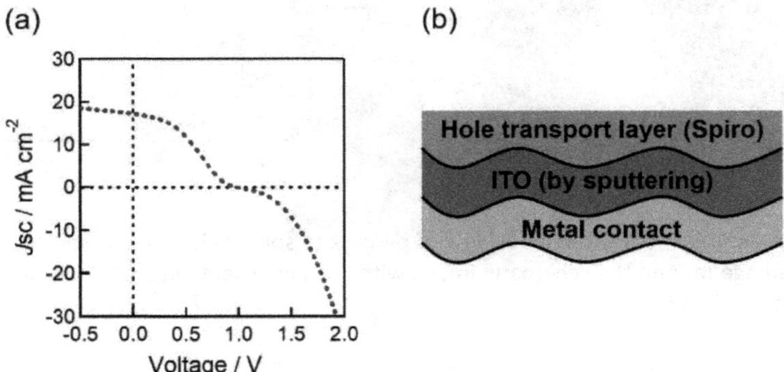

Figure 5.14 A representing photocurrent-voltage curve of the perovskite solar cell using a sputtered-ITO layer on an HTM (a) and the structure of sputtered-ITO and back-metal contact electrode on an HTM (b). The metal contact (Au) layers were deposited by thermal evaporation on the ITO layer in order to cancel the variation of conductivity between the ITO layers with the thickness variation (modified from Ref. [17]).

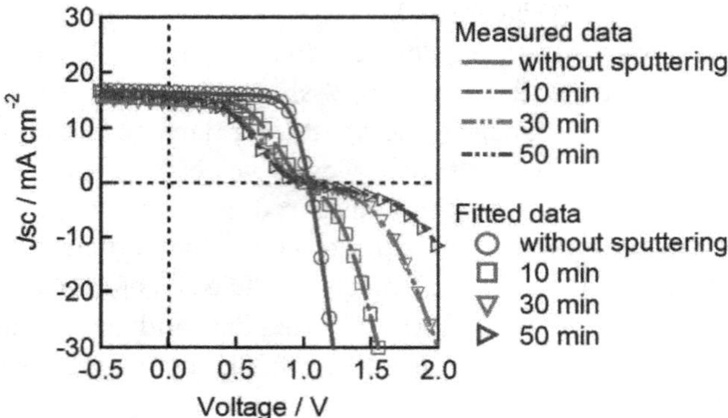

Figure 5.15 Photocurrent-voltage curves of perovskite solar cells (in Fig. 5.14) with changing ITO sputtering time (lines and plots show measured and fitted data, respectively) (modified from Ref. [39]).

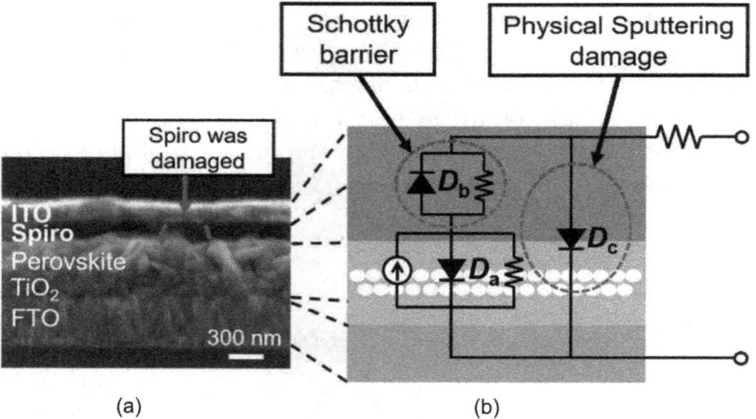

Figure 5.16 Cross-sectional view of perovskite solar cells with ITO sputtering damage (a) and the schematic image with the equivalent circuit (b) (modified from Ref. [39]).

5.6.2 Effectiveness of the Au Layer in Preventing ITO Sputtering Damage

To reduce ITO sputtering damage in the HTM layer, as discussed above, a protective layer was considered [17]. In this case, Au layers have been deposited on the surface of the HTM layer (Fig. 5.17). It was confirmed that the Au layer eliminates structural damage of the HTM layer and the Schottky diode barrier between ITO and the HTM, as shown in Fig. 5.18.

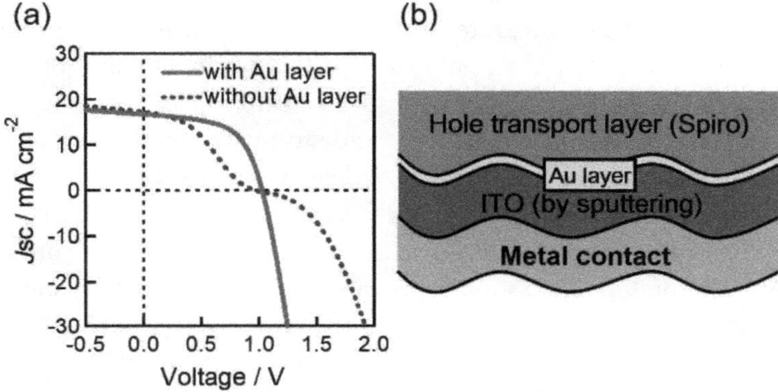

Figure 5.17 Photo *I–V* curves of perovskite solar cells with and without a Au layer between the HTM and ITO (a). The schematic image of <HTM/Au/ITO/back metal contact> interface (b) (modified from Ref. [17]).

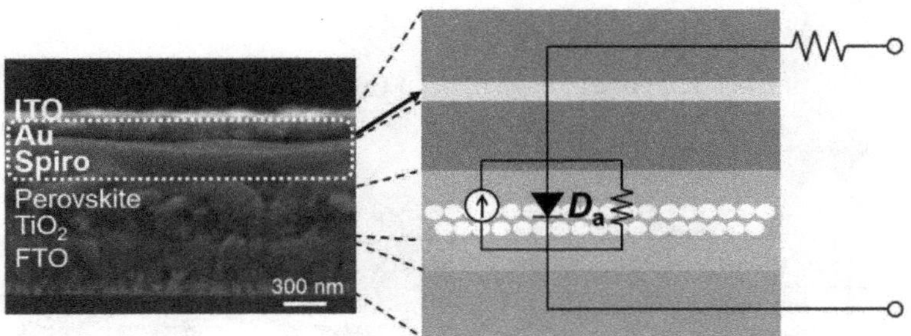

Figure 5.18 Cross-sectional SEM image (left) and the schematic image with the fitting equivalent circuit (tight) with a Au layer between the HTM (Spiro-OMeTAD) and ITO (modified from Ref. [39]).

5.6.3 Optoelectrical Engineering of Au and ITO Layer for a Tandem Solar Cell [17]

At present, the conversion efficiency of a two-terminal tandem device using a perovskite solar cell is lower than that of a single cell. Basically, since perovskite solar cells can be fabricated by low-temperature processing, the top perovskite layer is deposited on other bottom solar cells (silicon, CIGS, and perovskite). However, process optimization of the top perovskite layer is quite difficult due to the effect of the substrate (bottom cells) structure, which can be confirmed just after the finalization of cell fabrication and the PV measurements as tandem devices. If each cell (top and bottom) can be fabricated separately to be a tandem device, the optimization would be easier. Hence, we have developed the fabrication procedure of perovskite/silicon tandem solar cells by mechanical stacking for two- and four-terminal measurements. However, the contact between the top and bottom cells can be a point contact, as shown in Fig. 5.19, due to the roughness of the perovskite layer [17]. In this system, there is:

- A thin metal layer on the HTM of the perovskite solar cell for protection of the HTM against sputtering damage
- An ITO layer on the thin metal layer of the perovskite solar cell for the conducting electrode
- An ITO layer on the silicon solar cell as an antireflection coating (ARC)

The details have been described in the following sections.

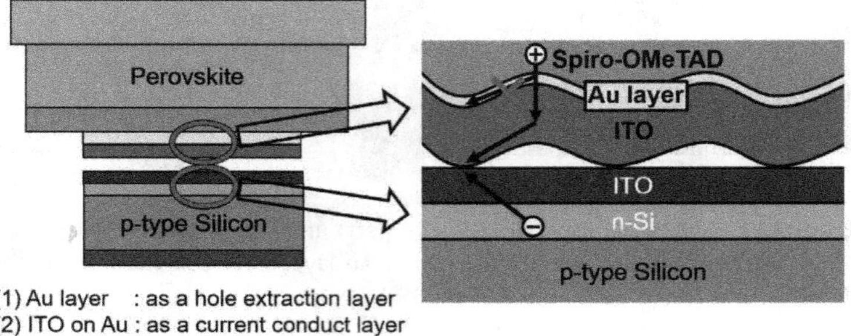

Figure 5.19 The structure of the perovskite/silicon tandem solar cell by mechanical stacking (left) and a schematic image of the point contact between perovskite and silicon mechanical-stacked tandem solar cells (modified from Ref. [17]).

5.6.3.1 Au or MoO_X between HTM and ITO layer

For a tandem device, the top perovskite solar cell needs light transparency in the region of the longer wavelength for the bottom silicon solar cells. Therefore, the thermally evaporated Au layer for ITO sputtering-damage protection needs to be thin enough to allow the light to pass. UV-Vis spectroscopy measurements using Au layers of different thicknesses show that a Au layer less than 2.5 nm has high transparency from 550 nm to 1100 nm, which shows no loss of photon for the bottom silicon solar cell (Fig. 5.20, Table 5.8) [17].

Another candidate to be used for protection against sputtering damage, MoOx, which can be deposited by thermal vacuum deposition, has been utilized between an HTM and sputtered IZO (Fig. 5.21). The resulting conversion efficiencies of transparent perovskite and two- and four-terminal tandem solar cells were 15.9%, 15.5%, and 19.8%. For conversion efficiency also, the optimal thickness of Au was 2.5 nm for the perovskite/silicon tandem cells [18].

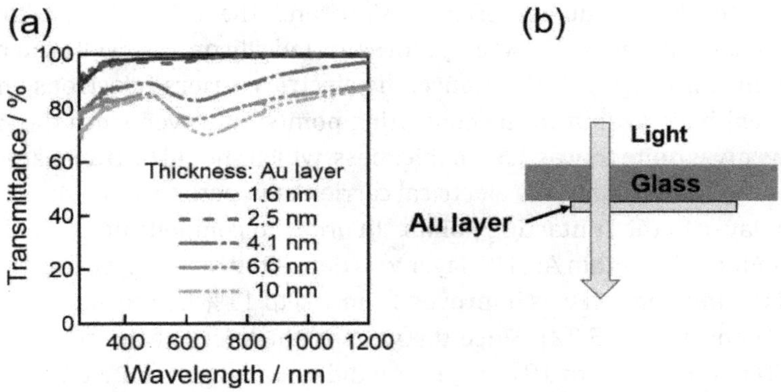

Figure 5.20 Transparency of Au layers of different thicknesses (a) and structure of a sample for light transmittance measurement (b) (modified from Ref. [17]).

Table 5.8 Average transparency of Au layers of different thicknesses [17]

Thickness of Au layer	1.6 nm	2.5 nm	4.1 nm	6.6 nm	10 nm
Average transmittance (550–1100 nm)	99.2%	99.2%	89.4%	81.7%	78.1%

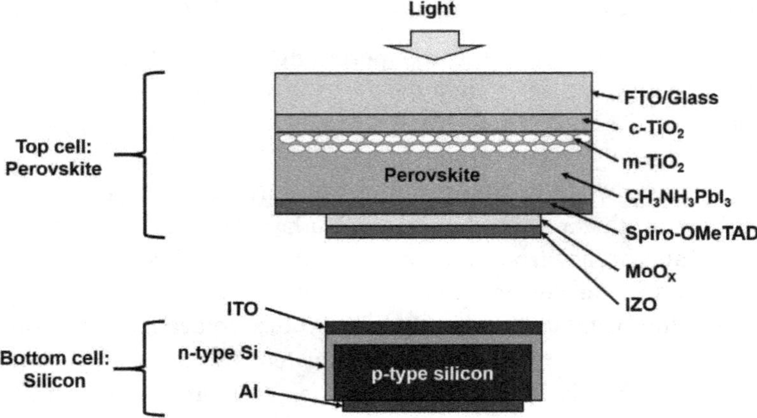

Figure 5.21 Structure of mechanical perovskite/silicon tandem cells using MoO*x* between the HTM and IZO (modified from Ref. [18]).

5.6.3.2 ITO layer on perovskite solar cells [17]

Due to the roughness of perovskite and the HTM surface, the mechanical contact between perovskite and silicon solar cells can be point contact (Fig. 5.19). Hence, the electric carriers (electrons and holes) have to shift to the contacting points. However, since the Au layer was quite thin as 2.5 nm thickness, which should be transparent for the bottom cells, the electrical carrier can't path through the thin Au layer to the contacting points. In order to compensate the small conductivity of thin Au, ITO layer was deposited on Au layer. The PCE of the tandem cell was improved from 1% to 11% by the addition of ITO on Au (Fig. 5.22). Since the varying thickness of the sputtered ITO (from 82 nm to 181 nm) on Au did not change the PV effects of tandem cells, the Au layer blocked the sputtering damage in the HTM quite effectively.

5.6.3.3 ITO layer on silicon solar cells [17]

It is necessary for the bottom silicon solar cells in a tandem device to absorb light pathed through the top perovskite layer. In the mechanical-stacked perovskite/silicon tandem cells, however, there is air in the gap between perovskite and silicon solar cells (Fig. 5.19). Without an ARC for the irradiation, the surface of silicon can reflect 40% of the incident light in the region of light from

550 nm to 1100 nm, which can be a significant optical loss (Fig. 5.23). As an ARC and conducting coating, an ITO layer was deposited on the silicon surface, which can reduce the average light reflection from 40% to 10% (Table 5.9). The resulting photonic current density was improved from 9.5 mA cm^{-2} to 12.5 mA cm^{-2} [17].

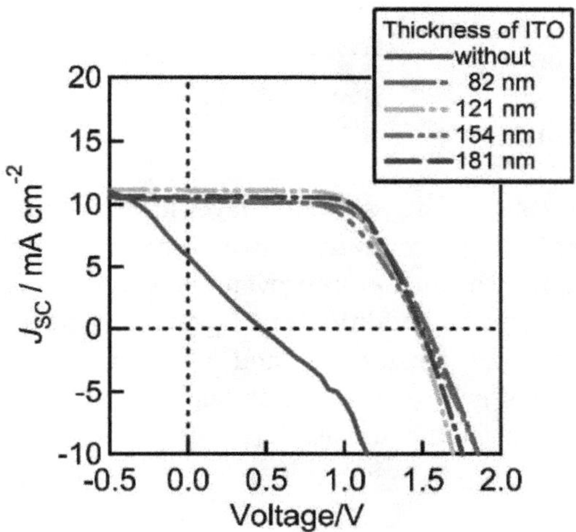

Figure 5.22 Photo *I–V* curves of two-terminal perovskite/silicon mechanical-stacked tandem solar cells with variation in the ITO thickness on the HTM of perovskite solar cells (modified from Ref. [17]).

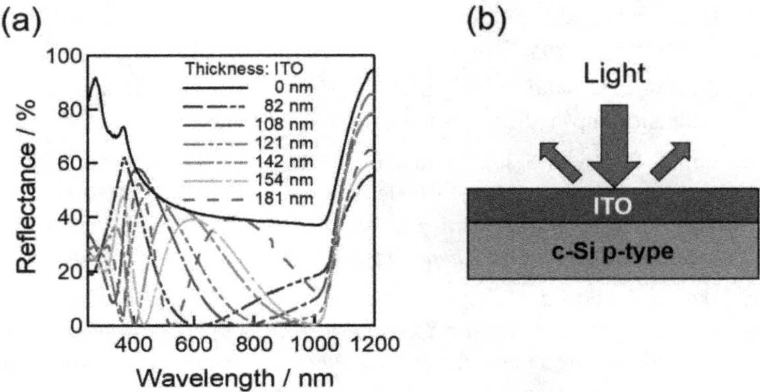

Figure 5.23 The reflection spectra of the silicon surface with variation in the thickness of the ITO layer on the silicon ware (a) and the schematic image of the ITO layer on a silicon substrate (b) (modified from Ref. [17]).

Table 5.9 Average reflection data on silicon with and without the ITO layer on silicon [17]

Thickness of ITO layer	0 nm	82 nm	108 nm	121 nm	142 nm	154 nm	181 nm
Average reflectance (550–1100 nm)	40.6%	11.2%	10.3%	11.4%	18.3%	19.8%	26.8%

5.7 Conclusion

Tandem solar cells using a perovskite layer have been improved with improvement in interface layers between top perovskite solar cells and bottom ones. For further improvement, bandgap engineering of the perovskite layer by addition of Br^- ions is required. Moreover, stability, large-area fabrication, and roll-to-roll processing are significant issues impeding commercialization, which need to be improved steadily for the future.

References

1. W. Shen, X. Chen, J. Qiu, J. A. Hayward, S. Sayeef, P. Osman, K. Meng, Z. Y. Dong (2020). A comprehensive review of variable renewable energy levelized cost of electricity. *Renew. Sust. Energ.*, **133**, 110301.
2. Green, M. A., Hishikawa, Y., Warta, W., Dunlop, E. D., Levi, D. H., Hohl-Ebinger, J., and Ho-Baillie, A. W. Y. (2017). Solar cell efficiency tables (version 50). *Prog. Photovoltaics Res. Appl.*, **25**, 668.
3. Shockley, W., and Queisser, H. J. (1961). Detailed balance limit of efficiency of p-n junction solar cells. *J. Appl. Phys.*, **32**, 510.
4. Löper, P., Moon, S.-J., Martín de Nicolas, S., Niesen, B., Ledinsky, M., Nicolay, S., Bailat, J., Yum, J.-H., De Wolf, S., and Ballif, C. (2015). Organic-inorganic halide perovskite/ crystalline silicon four-terminal tandem solar cells. *Phys. Chem. Chem. Phys.*, **17**, 1619.
5. Uzu, H., Ichikawa, M., Hino, M., Nakano, K., Meguro, T., Hernández, J. L., Kim, H. S., Park, N. G., and Yamamoto, K. (2015). High efficiency solar cells combining a perovskite and a silicon heterojunction solar cells via an optical splitting system. *Appl. Phys. Lett.*, **106**, 3.
6. Sheng, R., Ho-Baillie, A. W. Y., Huang, S., Keevers, M., Hao, X., Jiang, L., Cheng, Y.-B., and Green, M. A. (2015). Four-terminal tandem solar cells using $CH_3NH_3PbBr_3$ by spectrum splitting. *J. Phys. Chem. Lett.*, **6**, 3931.

7. Kinoshita, T., Nonomura, K., Jeon, N. J., Giordano, F., Abate, A., Uchida, S., Kubo, T., Seok, S., Nazeeruddin, M. K., Hagfeldt, A., Grätzel, M., and Segawa, H. (2015). Spectral splitting photovoltaics using perovskite and wideband dye-sensitized solar cells, *Nat. Commun.*, **6**, 8834.

8. Bailie, C. D., Christoforo, M. G., Mailoa, J. P., Bowring, A. R., Unger, E. L., Nguyen, W. H., Burschka, J., Pellet, N., Lee, J. Z., Grätzel, M., et al. (2015). Semi-transparent perovskite solar cells for tandems with silicon and CIGS. *Energy Environ. Sci.*, **8**, 956.

9. Duong, T., Lal, N., Grant, D., Jacobs, D., Zheng, P., Rahman, S., Shen, H., Stocks, M., Blakers, A., Weber, K., et al. (2016). Semitransparent perovskite solar cell with sputtered front and rear electrodes for a four- terminal tandem. *IEEE J. Photovoltaics*, **6**, 679.

10. Werner, J., Dubuis, G., Walter, A., Löper, P., Moon, S. J., Nicolay, S., Morales-Masis, M., De Wolf, S., Niesen, B., and Ballif, C. (2015). Sputtered rear electrode with broadband transparency for perovskite solar cells. *Sol. Energy Mater. Sol. Cells*, **141**, 407.

11. Werner, J., Barraud, L., Walter, A., Bräuninger, M., Sahli, F., Sacchetto, D., Tétreault, N., Paviet-Salomon, B., Moon, S.-J., Allebé, C., et al. (2016). Efficient near-infrared- transparent perovskite solar cells enabling direct comparison of 4-terminal and monolithic perovskite/silicon tandem cells. *ACS Energy Lett.*, **1**, 474.

12. Bush, K. A., Bailie, C. D., Chen, Y., Bowring, A. R., Wang, W., Ma, W., Leijtens, T., Moghadam, F., and McGehee, M. D. (2016). Thermal and environmental stability of semi- transparent perovskite solar cells for tandems enabled by a solution-processed nanoparticle buffer layer and sputtered ITO Electrode. *Adv. Mater.*, **28**, 3937.

13. Chen, B., Bai, Y., Yu, Z., Li, T., Zheng, X., Dong, Q., Shen, L., Boccard, M., Gruverman, A., Holman, Z., et al. (2016). Efficient semitransparent perovskite solar cells for 23.0%-efficiency perovskite/silicon four-terminal tandem cells. *Adv. Energy Mater.*, **6**, 1601128.

14. McMeekin, D. P., Sadoughi, G., Rehman, W., Eperon, G. E., Saliba, M., Horantner, M. T., Haghighirad, A., Sakai, N., Korte, L., et al. (2016). A mixed-cation lead mixed-halide perovskite absorber for tandem solar cells. *Science*, **351**, 151.

15. Duong, T., Wu, Y., Shen, H., Peng, J., Fu, X., Jacobs, D., Wang, E. C., Kho, T. C., Fong, K. C., Stocks, M., et al. (2017). Rubidium multication perovskite with optimized bandgap for perovskite-silicon tandem with over 26% efficiency. *Adv. Energy Mater.*, **7**, 1700228.

16. Lang, F., Gluba, M. A., Albrecht, S., Rappich, J., Korte, L., Rech, B., and Nickel, N. H. (2015). Perovskite solar cells with large-area CVD-graphene for tandem solar cells. *J. Phys. Chem. Lett.*, **6**, 2745.

17. Kanda, H., Uzum, A., Nishino, H., Umeyama, T., Imahori, H., Ishikawa, Y., Uraoka, Y., and Ito, S. (2016). Interface optoelectronics engineering for mechanically stacked tandem solar cells based on perovskite and silicon. *ACS Appl. Mater. Interfaces*, **8**, 33553.

18. Kanda, H., Shibayama, N., Uzum, A., Umeyama, T., Imahori, H., Chiang, Y.-H., Chen, P., Nazeeruddin, M. K., and Ito, S. (2018). Facile fabrication method of small-sized crystal silicon solar cells for ubiquitous applications and tandem device with perovskite solar cells. *Mater. Today Energy*, **7**, 190–198.

19. Mailoa, J. P., Bailie, C. D., Johlin, E. C., Hoke, E. T., Akey, A. J., Nguyen, W. H., McGehee, M. D., and Buonassisi, T. (2015). A 2-terminal perovskite/silicon multijunction solar cell enabled by a silicon tunnel junction. *Appl. Phys. Lett.*, **106**, 121105.

20. Albrecht, S., Saliba, M., Correa-Baena, J. P., Lang, F., Kegelmann, L., Mews, M., Steier, L., Abate, A., Rappich, J., Korte, L., et al. (2016). Monolithic perovskite/silicon-heterojunction tandem solar cells processed at low temperature. *Energy Environ. Sci.*, **9**, 81.

21. Werner, J., Weng, C. H., Walter, A., Fesquet, L., Seif, J. P., De Wolf, S., Niesen, B., and Ballif, C. (2016). Efficient monolithic perovskite/ silicon tandem solar cell with cell area >1 cm^2. *J. Phys. Chem. Lett.*, **7**, 161.

22. Werner, J., Barraud, L., Walter, A., Bräuninger, M., Sahli, F., Sacchetto, D., Tétreault, N., Paviet-Salomon, B., Moon, S.-J., Allebé, C., et al. (2016). Efficient near-infrared- transparent perovskite solar cells enabling direct comparison of 4-terminal and monolithic perovskite/silicon tandem cells. *ACS Energy Lett.*, **1**, 474.

23. Bush, K. A., Palmstrom, A. F., Yu, Z. J., Boccard, M., Cheacharoen, R., Mailoa, J. P., McMeekin, D. P., Hoye, R. L. Z., Bailie, C. D., Leijtens, T., et al. (2017). 23.6%-Efficient monolithic perovskite/silicon tandem solar cells with improved stability. *Nat. Energy*, **2**, 17009.

24. Lee, M., Park, S. J., Hwang, Y. J., Jun, Y., and Min, B. K. (2016). Tandem architecture of perovskite and $Cu(In,Ga)(S,Se)_2$ created by solution processes for solar cells. *Adv. Opt. Mater.*, **4**, 2102.

25. Kranz, L., Abate, A., Feurer, T., Fu, F., Avancini, E., Löckinger, J., Reinhard, P., Zakeeruddin, S. M., Grätzel, M., Buecheler, S., et al. (2015). High-efficiency polycrystalline thin film tandem solar cells. *J. Phys. Chem. Lett.*, **6**, 2676.

26. Yang, Y. M., Chen, Q., Hsieh, Y.-T., Song, T.-B., De Marco, N., Zhou, H., and Yang, Y. (2015). Multilayer transparent top electrode for solution processed perovskite/ Cu(In,Ga)(Se,S)$_2$ four terminal tandem solar cells. *ACS Nano*, **9**, 7714.

27. Fu, F., Feurer, T., Jäger, T., Avancini, E., Bissig, B., Yoon, S., Buecheler, S., and Tiwari, A. N. (2015). Low-temperature-processed efficient semi-transparent planar perovskite solar cells for bifacial and tandem applications. *Nat. Commun.*, **6**, 8932.

28. Fu, F., Feurer, T., Weiss, T. P., Pisoni, S., Avancini, E., Andres, C., Buecheler, S., and Tiwari, A. N. (2016). High-efficiency inverted semi-transparent planar perovskite solar cells in substrate configuration. *Nat. Energy*, **2**, 16190.

29. Guchhait, A., Dewi, H. A., Leow, S. W., Wang, H., Han, G., Suhaimi, F. B., Mhaisalkar, S., Wong, L. H., and Mathews, N. (2017). Over 20% efficient CIGS-perovskite tandem solar cells. *ACS Energy Lett.*, **2**, 807.

30. Todorov, T., Gershon, T., Gunawan, O., Lee, Y. S., Sturdevant, C., Chang, L. Y., and Guha, S. (2015). Monolithic perovskite-CIGS tandem solar cells via in situ band gap engineering. *Adv. Energy Mater.*, **5**, 1500799.

31. Yang, Z., Rajagopal, A., Chueh, C. C., Jo, S. B., Liu, B., Zhao, T., and Jen, A. K. Y. (2016). Stable low-bandgap Pb-Sn binary perovskites for tandem solar cells. *Adv. Mater.*, **28**, 8990.

32. Eperon, G. E., Leijtens, T., Bush, K. A., Prasanna, R., Green, T., Wang, J. T.-W., Mc Meekin, D. P., Volonakis, G., Milot, R. L., May, R., et al. (2016). Perovskite-perovskite tandem photovoltaics with optimized band gaps. *Science*, **354**, 861.

33. Zhao, D., Yu, Y., Wang, C., Liao, W., Shrestha, N., Grice, C. R., Cimaroli, A. J., Guan, L., Ellingson, R. J., Zhu, K., et al. (2017). Low-bandgap mixed tin-lead iodide perovskite absorbers with long carrier lifetimes for all-perovskite tandem solar cells. *Nat. Energy*, **2**, 17018.

34. Heo, J. H., and Im, S. H. (2016). $CH_3NH_3PbBr_3$-$CH_3NH_3PbI_3$ perovskite-perovskite tandem solar cells with exceeding 2.2 V open circuit voltage. *Adv. Mater.*, **28**, 5121.

35. Jiang, F., Liu, T., Luo, B., Tong, J., Qin, F., Xiong, S., Li, Z., and Zhou, Y. (2016). A two- terminal perovskite/perovskite tandem solar cell. *J. Mater. Chem. A*, **4**, 1208.

36. Forgács, D., Gil-Escrig, L., Pérez-del-Rey, D., Momblona, C., Werner, J., Niesen, B., Ballif, C., Sessolo, M., and Bolink, H. J. (2017). Efficient monolithic perovskite/perovskite tandem solar cells. *Adv. Energy. Mater.*, **7**, 1602121.

37. Rajagopal, A., Yang, Z., Jo, S. B., Braly, I. L., Liang, P.-W., Hillhouse, H. W., and Jen, A. K.-Y. (2017). Highly efficient perovskite-perovskite tandem solar cells reaching 80% of the theoretical limit in photovoltage. *Adv. Mater.*, **29**, 1702140.

38. Ito, S., Kanaya, S., Nishino, H., Umeyama, T., and Imahori, H. (2015). Material exchange property of organo lead halide perovskite with hole-transporting materials. *Photonics*, **2**, 1043.

39. Kanda, H., Uzum, A., Baranwal, A. K., Peiris, T. A. N., Umeyama, T., Imahori, H., Segawa, H., Miyasaka, T., and Ito, S. (2016). Analysis of sputtering damage on I-V curves for perovskite solar cells and simulation with reversed diode model. *J. Phys. Chem. C*, **120**, 28441.

Chapter 6

Perovskite Resistive Memory

Bohee Hwang, Youngjun Park, and Jang-Sik Lee
Department of Materials Science and Engineering,
Pohang University of Science and Technology (POSTECH),
Pohang 37673, Korea
jangsik@postech.ac.kr

6.1 Introduction

Recently, organic–inorganic hybrid perovskites (OIPs) with a chemical formula of ABX_3, where A = monovalent cation (Cs^+, $CH_3NH_3^+$), B = bivalent metal cation (Pb^{2+}, Sn^{2+}), and X = halide anion (I^-, Br^-, Cl^-), have rapidly attracted a lot of attention due to their superior photovoltaic properties, such as high carrier mobility, long electron-hole diffusion length, and strong photoluminescence (PL) [1–4]. These properties enable their application in solar cells, light-emitting diodes, and photodetectors [5–8]. Perovskite photovoltaic devices show a difference in current (hysteresis) under forward and reverse scanning directions. Also, the photocurrent direction in $CH_3NH_3PbI_3$ films can be flipped by applying a small electric field $<1\ V\ \mu m^{-1}$; this trait is a result of reversible p-i-n junctions, which are induced by a defect drift in the perovskite layer under an electric

Multifunctional Organic–Inorganic Halide Perovskite: Applications in Solar Cells, Light-Emitting Diodes, and Resistive Memory
Edited by Nam-Gyu Park and Hiroshi Segawa

ISBN 978-981-4800-52-5 (Hardcover), 978-1-003-27593-0 (eBook)
www.jennystanford.com

field [9]. These behaviors may be a result of ion migration; this effect can be exploited to develop memory devices, such as resistive switching memory.

Resistive random-access memory (ReRAM) has been considered to be the most important candidate for the next generation of nonvolatile devices because ReRAM has fast switching speed, low power consumption, and high scalability [10, 11]. A ReRAM device is a two- terminal structure with an insulating layer sandwiched between two conducting electrodes. An external electric field can switch the memory cell into a low-resistance state (LRS, on state) and a high-resistance state (HRS, off state) [12]. ReRAM is classified into unipolar and bipolar switching modes. In unipolar switching, the polarity of the switching voltage is irrelevant, whereas in bipolar switching the electrical polarity required to switch from an HRS to an LRS is opposite to that required to switch from an LRS to an HRS. To obtain ReRAM that has reliable switching, a high on/off current ratio, and long retention time, various materials have been evaluated as switching materials, including organics, binary oxides, and perovskite oxides [13–15]. Especially, OIP materials offer advantages for application in ReRAM, such as flexibility, multilevel property, and analog switching [16–34].

This chapter discusses the integration of different OIP materials into nonvolatile memory devices and explains their characterization and mechanisms for ReRAM applications. First, we will introduce ReRAM in terms of operation mechanism, requirements, etc. This chapter focuses on various types of OIP-based ReRAM and considers their systems, properties, and memory characteristics in detail. We will define the hysteretic behavior and then examine the microscopic model focusing ion migration. We cover recent advances in OIP-based ReRAM, including flexible ReRAM based on OIP, an approach to increase the long-term stability of these ReRAM devices, analog switching property that could be applied to neuromorphic applications, and high-density memory applications. Finally, we will give a brief outlook on future directions for OIP-based ReRAM.

6.2 Resistive Switching Memory

Resistive switching memory is composed of a capacitor-like metal-insulator-metal (MIM) structure, and the information is stored as different resistance states, such as LRS and HRS. Application of an external electric field can change these states reversibly. The switch process from an HRS to an LRS is called a "set," and the switch from an LRS to an HRS is called a "reset" [35]. The resistive switching mechanism involves a redox reaction. The switching mechanism can be considered as the filamentary- and interface-type. Filamentary-type ReRAM can be classified as electrochemical metallization memory (ECM) and valence change memory (VCM). ECM, also known as conductive bridging random-access memory, depends on an electrochemically active metal electrode, such as Ag or Cu. This switching mechanism uses an active metal top electrode in the MIM structure [36]. First, mobile metal cations drift through the insulating layer. Then, cations that have migrated from the top electrode are reduced to metal at the bottom electrode (inert counterelectrode); the result is a highly conductive metal filament in the insulator. In VCM, the conductive filament is generated by oxygen vacancy defects instead of metal cations and anion diffusion that occurs within the insulation layer. Most of the VCM effect happens at the transition metal oxide or perovskite oxides, and the conductive filaments are formed by electrons trapped in oxygen vacancies [10, 37]. Especially, oxygen ions have an important function in the switching mechanism. For example, the generation of oxygen vacancies and the migration of defects from the transition metal oxide lattice result in the formation of conductive filaments [35]. Interface-type memory is governed by a field-induced change of Schottky barriers at the interface between electrode and insulating layer [12, 36]. A change of the barrier height between the anode and the insulating layer can lead to the resistance switching on the basis of a redox reaction.

6.3 Origin of the Hysteresis Ion/Defect Migration

Hysteresis in OIP materials prevent perovskite solar cells from large-scale commercial applications due to a difference in efficiency with sweeping direction. These effects have garnered a lot of interest in the research field [38–48]. Hysteresis in the *I–V* curves of OIP materials means that the photocurrent response under forward bias differs from that under reverse bias (Fig. 6.1a). Three mechanisms have been proposed to explain hysteresis: charge trapping [49], ferroelectricity [50], and ion migration [51]. In this chapter, we will focus on ion migration and ionic species that are mobile in the perovskite film because these concepts can help to explain the resistive switching effect.

To investigate the ion migration effect in perovskite film, the ion that migrates must be identified. We focus on methyl ammonium lead iodide films. They include several defects, such as vacancies (iodine vacancy V_I, methyl ammonium vacancy V_{MA}, and lead vacancy V_{Pb}), interstitials (MA_i, Pb_i, and I_i), and antisite substitutions (MA_I and Pb_I). Due to high activation energy (E_A = ~0.80 eV) [52] and a low migration rate, V_{Pb} has not been considered in the resistive switching effect in previous studies. MA^+ ions are considered to be the major reason for the giant field switchable photovoltaic effect in perovskite films [9, 53]. However, the estimated diffusion coefficient of I^- is 10^{-12} cm^2 s^{-1}, which is 4 orders of magnitude greater than that of MA^+ (10^{-16} cm^{2}s^{-1}) [54]. This result indicates that hysteresis may be caused more by I^- ions than by MA^+ ions.

Ion migration in a solid material can be determined by the activation energy (E_A), and the ion migration rate (r_m) can be affected by E_A, $r_m \propto \exp\left(\frac{-E_A}{K_B T}\right)$, where K_B = 8.617 × 10^{-5} eV/K is the Boltzmann constant and T [K] is the absolute temperature [55]. E_A depends on the ionic species, so determination of E_A helps to identify the ions that move through the film. To obtain detailed understanding of the migration process in perovskites, theoretical density functional theory (DFT) calculations were carried out, which determined that I^- has E_A = 0.58 eV, MA^+ has E_A = 0.84 eV, and Pb^{2+} has E_A = 2.31 eV. The model suggests that V_I migrates along the octahedron edge, which is the shortest pathway in the perovskite structure (Fig. 6.1b), and this leads to the low E_A of I^- [54]. However, the MA^+ diffuses in the (100) plane along the <100> direction so this ion's E_A is higher than that of I^-. Pb^{2+} moves in the diagonal direction (Fig. 6.1c, d). Another theoretical work calculated E_A = 0.08 eV for I^-, E_A = 0.46 eV for MA^+, and E_A = 0.08 eV for Pb^{2+} [52]. Although the two calculation methods obtained different E_A values, they agreed that I^- moves faster than MA^+ and Pb^{2+}. Therefore, we can consider that an OIP materials can be used in other electronics that exploit the effect of mobile ions or defects such as I^-.

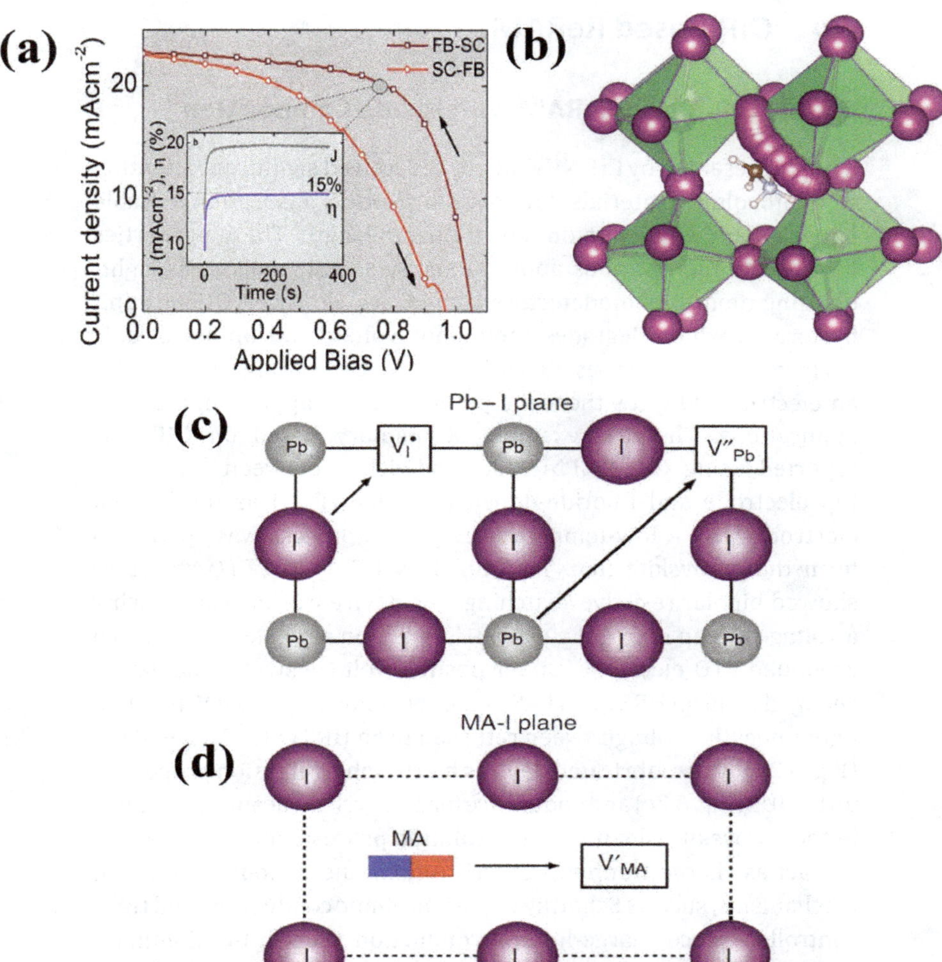

Figure 6.1 (a) Forward bias to short circuit and short circuit to forward bias current-voltage curves measured under AM 1.5 simulated sunlight. Reprinted with permission from Ref. [48], Copyright 2014 American Chemical Society. (b) Calculated path indicating the slight curvature and local relaxation/tilting of the octahedral. Schematic diagram of ionic migration, especially I^-, Pb^{2+}, and MA^+: (c) I^- ions move along the octahedron edge, Pb^{2+} moves along the diagonal direction, and (d) MA^+ moves via neighboring vacancy sites. Reproduced with permission from Ref. [54], Copyright 2015 Nature Publishing Group.

6.4 OIP-Based ReRAM

6.4.1 OIP-Based ReRAM with Halide Composition

OIPs, represented by $CH_3NH_3PbI_3$, have received significant attention as photovoltaic materials due to strong optical absorption, tunable bandgap, and long electron-hole diffusion length. These properties have led to their various applications, such as in solar cells, light-emitting diodes, photodetectors, and lasers. However, OIPs exhibit hysteresis, which degrades their photovoltaic and optoelectronic performance. Hysteresis in OIPs due to defect migration under an electric field gives them the potential to be applied in memory applications. The first example of memory based on OIP was reported using $CH_3NH_3PbI_{3-x}Cl_x$ sandwiched between Au as the top electrode and fluoride-doped tin oxide (FTO) as the bottom electrode [16]. A low-temperature solution process was applied to form the perovskite film. The Au/$CH_3NH_3PbI_{3-x}Cl_x$/FTO structure showed bipolar resistive switching. The device was measured using a voltage sweep of $0 \rightarrow 1 \rightarrow 0 \rightarrow -1 \rightarrow 0$ V on a Au electrode with a grounded FTO electrode. Under positive voltage sweep, the device changed from an HRS to an LRS at the set voltage V_{set} ~0.8 V and then under negative voltage sweep returned to an HRS with V_{reset} ~−0.6 V (Fig. 6.2a,b). The fabricated devices had reliable retention properties up to 10^4 s (Fig. 6.2c) and endurance for 100 cycles. Vacancies can form in the perovskite film during the solution process, and these defects can act as charge trapping centers. Among the various conduction mechanisms, such as Schottky emission, ohmic conduction, and trap-controlled space-charge-limited conduction (SCLC), the dominant conduction mechanism is trap-controlled SCLC due to the defects in the perovskite film. Under positive bias, the I–V curve follows Ohm's law at a low voltage (0 to 0.4 V), mainly Poole–Frenkel conduction in the middle region (0.4 to 0.5 V), and Child's law at a high voltage (>0.5 V). At a low voltage the electric field that is applied to the device is not sufficient to drive the movement of charge carriers, so ohmic conduction occurs; this process produces more thermally generated free charge carriers than the number of injected carriers. As the applied voltage increases, traps in $CH_3NH_3PbI_{3-x}Cl_x$ materials become occupied with injected carriers, so current I increases in proportion to V^2.

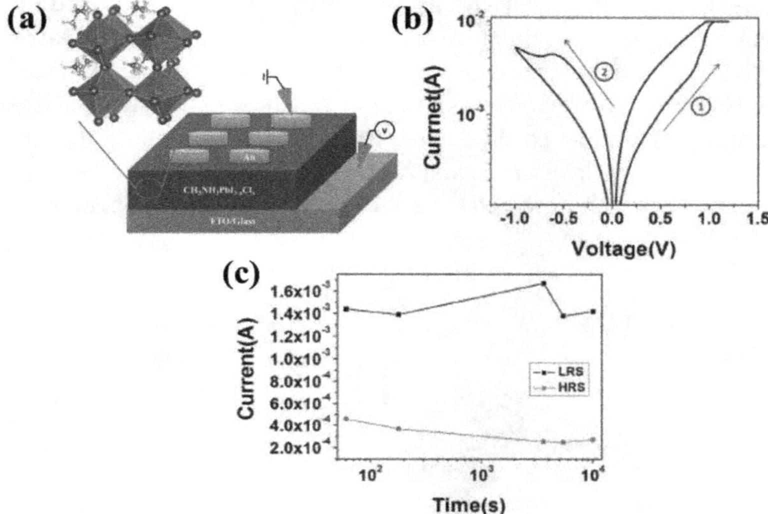

Figure 6.2 (a) Schematic image of the fabricated Au/$CH_3NH_3PbI_{3-x}Cl_x$/FTO device. Resistive switching properties of the Au/$CH_3NH_3PbI_{3-x}Cl_x$/FTO device. (b) *I–V* curves of the device and (c) retention properties of the device. Reprinted from Ref. [16], Copyright 2015, with permission of John Wiley & Sons.

$MAPbI_{3-x}Br_x$ has been exploited to fabricate a memory device [20]. The $MAPbI_{3-x}Br_x$ ($x = 0, 1, 2, 3$) film was formed by solvent engineering, which produced a homogenous film quality. The thickness of the perovskite layer changed with the halide content, so the electric field F_E was used to indicate the memory property instead of the applied bias. The I-F_E curve showed bipolar resistive switching behavior. When the applied F_E was swept from 0 to 9.41×10^4 V/cm, the resistance changed from an HRS to an LRS; after a negative F_E was applied, the resistance returned to an HRS (Fig. 6.3a,b). Among the four compositions of halide perovskites, $MAPbBr_3$ showed the lowest set F_E (Fig. 6.3c); this sensitivity suggests that the ions or defects in $MAPbBr_3$ may be highly mobile.

Because the barrier to the migration of halide vacancies is low, they migrate along the shortest path along the edges of the octahedral. Therefore, we suggest that resistive switching occurs by the formation of conductive filaments that are related to halide vacancies and to charge trapping and detrapping induced by an electric field. When a sufficiently positive voltage is applied to the top electrode, positively charged halide vacancies move toward the bottom electrode (indium tin oxide [ITO]). As the voltage increases, filaments of halide vacancies may form from the bottom electrode to the top electrode. When the opposite voltage is applied, charge detrapping occurs, so the filament ruptures. Overall, $MAPbI_{3-x}Br_x$-based memory devices showed reliable endurance under AC voltage pulses. Data were retained for 2×10^4 s with a constant on/off ratio of $\sim 10^2$.

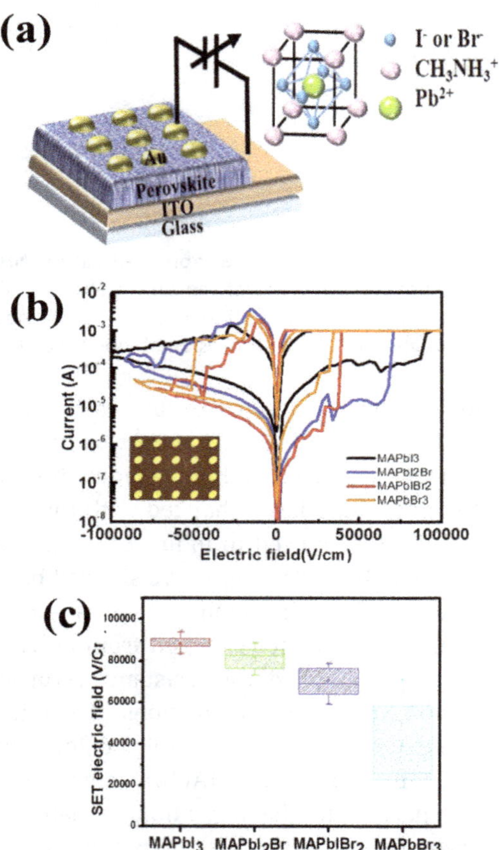

Figure 6.3 (a) Schematic diagram of a memory device with a Au/hybrid perovskite/ITO/glass substrate and the structure of perovskite. (b) *I–V* characteristic of $Au/CH_3NH_3PbI_{3-x}Br_x/ITO$. (c) Statistical distribution of set electric fields of hybrid perovskite resistive switching memory. Reproduced with permission from Ref. [20], Copyright 2017 Nature Publishing Group.

6.4.2 Flexible OIP-Based ReRAM

Some future electronics, such as sensors and displays, will be foldable and stretchable. Flexible electronics require flexible circuit components, particularly the nonvolatile memory devices.

A flexible ReRAM with Au/$CH_3NH_3PbI_3$/ITO/polyethylene terephthalate (PET) showed bipolar resistive switching memory [17]. To obtain a homogeneous perovskite layer, the one-step toluene dripping method can be used, which quickly crystallizes the perovskite film; the process yields a smooth surface. The *I–V* curve of the perovskite-ReRAM showed resistive switching properties. Application of positive bias switched the device at V_{set} ~0.7 V, and a reverse bias sweep returned the device to an HRS (Fig. 6.4a). Flexible perovskite ReRAM on PET was stable under tensile and compressive stress to a radius of curvature r_c = 1.5 cm; even at r_c = 7.5 mm, the electrical characteristics under bending states did not change much (Fig. 6.4b). Moreover, the device maintained its switching capability up to 100 bending cycles (Fig. 6.4c). A double logarithmic plot of the *I–V* curve showed that the conduction mechanism was SCLC at an HRS and ohmic at an LRS. SCLC indicates that charge trapping sites may form in the $CH_3NH_3PbI_3$ layer. The resistive switching mechanism of $CH_3NH_3PbI_3$ may be a result of vacancy migration and charge trapping under an electric field. Perovskite layers can include many types of defects, such as vacancies and interstitials. Among the defects, V_I has the lowest E_A, of 0.58 eV [54]; therefore iodine vacancies were the main cause of resistive switching. As the positive bias increases, positively charged iodine vacancies move toward the negatively charged ITO during the set process. As the voltage increases, filaments of iodine vacancies connect the bottom electrode to the top electrode. During the reset process, electron detrapping occurs, so the conductive filaments are ruptured. $CH_3NH_3PbBr_3$ quantum dots have been also used in flexible nonvolatile memory applications [19]. The perovskite quantum dot (PeQD) solution was fabricated by the solution process assisted by an organic capping ligand. The colloidal PeQD solution was blended in polymethylmethacrylate (PMMA), and the resistive switching layer was formed by spin coating.

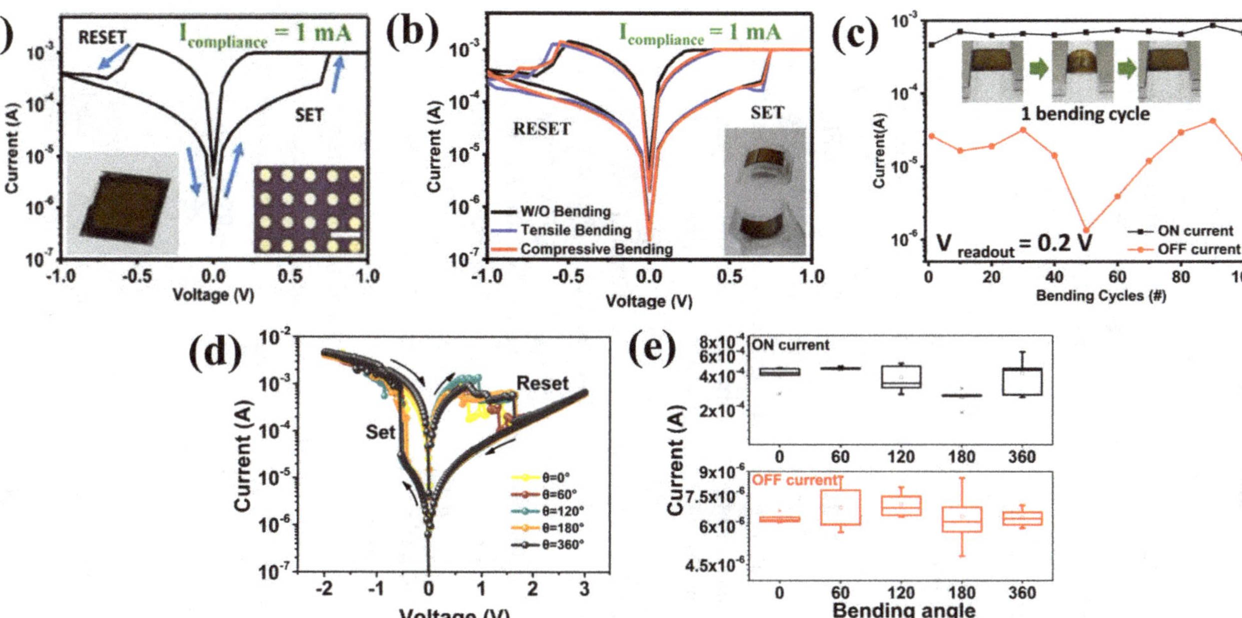

Figure 6.4 (a) *I–V* characteristic of the Au/perovskite/ITO structure. Mechanical flexibility of a hybrid OIP-based memory device: (b) *I–V* curves without and with bending to a radius of 1.5 cm. The inset shows tensile and compressive bending states of devices. (c) Bending stability of perovskite ReRAM with repetitive bending cycles. The inset shows the process of one bending cycle. Reprinted with permission from Ref. [17], Copyright (2016) American Chemical Society. (d) *I–V* curve of an Al/$CsPbBr_3$/PEDOT:PSS/ITO device. (e) Bending stability at different bending angles of perovskite resistive switching memory with repetitive bending cycles. Reprinted with permission from Ref. [18], Copyright (2014) American Chemical Society.

An OIP-based ReRAM with Ag/PMMA/PeQDs:PMMA/PMMA/ITO has been fabricated, and the *I*–*V* curve showed bipolar resistive switching behavior. When an applied bias was swept from 0 to 2 V, the resistance changed from an HRS to an LRS at 1 V. The ReRAM had a bipolar property, so the LRS was maintained until a voltage of different polarity was applied to change the device from an LRS to an HRS [19]. The PeQD-based ReRAM was fabricated on a PET substrate to confirm the feasibility of flexible electronics. The on/off ratio (>10^3) declined slightly after 10 times bending to r_c = 7 mm.

$CsPbBr_3$-based flexible ReRAM has been fabricated in an MIM structure of Al/$CsPbBr_3$/PEDOT:PSS/ITO/PET [18]. The $CsPbBr_3$ active layer was deposited using a two-step sequential deposition technique. The *I*–*V* curves showed bipolar resistive switching. The device was measured using a voltage sequence of 0 → -2 → 0 → 3 → 0 V on an Al electrode with a grounded ITO electrode. The electroforming voltage was formed at 3 V, and a set voltage V_{set} ~–0.6 V and reset voltage V_{reset} ~1.7 V were observed. The device also showed self-compliance behavior.

To check the feasibility of this ReRAM in flexible electronics, different angles (0° → 60° → 120° → 180° → 360°) of bending were applied to the device for 100 cycles, and the device retained the on/off current ratio (about 10^2) after the bending cycles (Fig. 6.4d,e).

6.4.3 Possibility of High-Density Memory Applications: Multilevel and Nanoscale Memory

The demand for increased data storage is an important challenge. High memory capacity using resistive switching memory can be achieved by multilevel switching or downscaling of device size. To extend its application to high-density information storage, OIP-based ReRAM that has multilevel property has been developed. A thin film of $CH_3NH_3PbI_3$ was sandwiched between Ag as the top electrode and Pt as the bottom electrode (Fig. 6.5a) [34]. The memory device switched from an HRS to an LRS under positive bias sweep with V_{set} ~0.13 V, then changed back to an HRS under negative bias sweep; on/off ratio was larger than 10^6. Especially, this device exhibited a low set electric field of 3.25 × 10^3 V/cm, which is lower than that of other inorganic perovskites. To confirm the electrical reliability of the device, retention property maintained 11000 s and switching endurance was stable up to 350 cycles. The Ag/$CH_3NH_3PbI_3$/Pt/Ti/SiO_2 structure demonstrated a high on/off ratio (>10^6)—this property has the possibility to be exploited for multilevel information storage. This property was achieved by tuning the compliance current at 10^{-2}, 10^{-4}, 10^{-5}, and 10^{-6} A during the set process; four distinct LRS levels were achieved (Fig. 6.5b,c). The origin of the resistive switching effect of $CH_3NH_3PbI_3$ was explained by the formation of conductive filaments by I_i' and ruptured by $V^{\bullet}_I$.

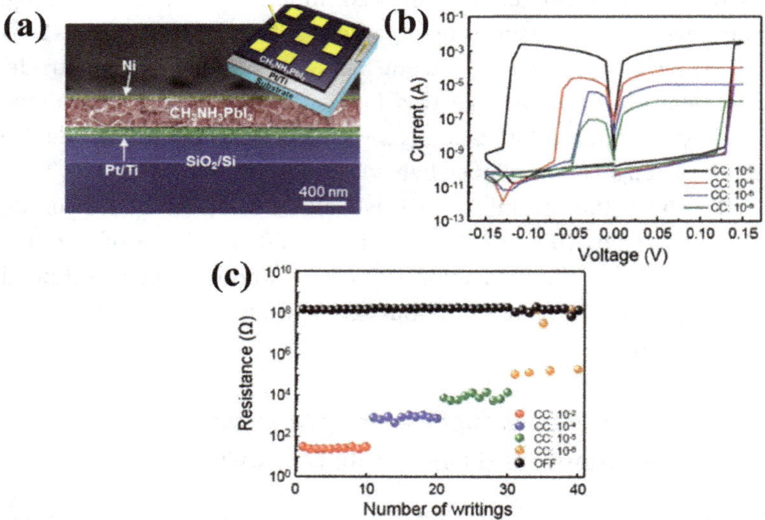

Figure 6.5 (a) Cross-sectional SEM image of a Ni/$CH_3NH_3PbI_3$/Pt/Ti/SiO_2/Si structure and schematic image of the metal/$CH_3NH_3PbI_3$/metal vertical structure for resistive switching (inset). Multilevel resistive switching properties: (b) *I–V* curves of Ag/$CH_3NH_3PbI_3$/Pt cells under compliance currents (CC) = 10^{-2}, 10^{-4}, 10^{-5}, and 10^{-6} A. (c) Reversible resistive switching over 40 cycles with different current compliances of 10^{-2}, 10^{-4}, 10^{-5}, and 10^{-6} A. Reprinted from Ref. [34], Copyright 2016, with permission of John Wiley & Sons.

To further increase the density of information storage, memory cells should be stacked in 3D arrays or the device size should be scaled down. However, the existing solution process to fabricate organolead halide perovskite cannot be easily adapted to deposit a nanometer-scale template. To evade this limitation, sequential vapor deposition has been used to deposit OIP-based resistive switching layers in nanoscale holes on a silicon wafer [23]. This method enables the use of organolead halide perovskites in practical large-scale memory devices that are compatible with complementary metal-organic semiconductor (CMOS) technology. This approach produced successful memory properties, such as long data retention ($>10^5$ s), good endurance (500 cycles), and fast switching (200 ns) even in very small device sizes. This work on preparing $CH_3NH_3PbI_3$-based nanoscale devices and the cross-point array structure may be an important step for high-capacity information storage.

6.4.4 Air-Stable OIP-Based ReRAM

Despite the recent progress in OIP materials, the instability of $CH_3NH_3PbI_3$ under humid and ambient conditions prevents its commercialization. To increase the long-term stability of OIP materials, a metal-oxide layer can be used to protect the underlying perovskite film. Resistive switching memory that incorporates $CH_3NH_3PbI_3$ as a switchable material sandwiched between Au and ITO has been demonstrated [21]. To protect the perovskite layer, a layer of metal oxide (e.g., ZnO or AlO_x) was deposited on top of the $CH_3NH_3PbI_3$ layer. The memory devices with Au/$CH_3NH_3PbI_3$/ITO and Au/ZnO/$CH_3NH_3PbI_3$/ITO showed bipolar nonvolatile resistive switching behavior [21]. The devices were measured under a voltage sweep of $0 \rightarrow 2 \rightarrow 0 \rightarrow -1.5 \rightarrow 0$ V on a Au electrode with a grounded ITO electrode. Both devices had set voltages of 0.9 V and 1.1 V, respectively. The device stability was measured under ambient atmosphere. The memory device that was encapsulated by ZnO maintained constant on/off ratio for up to 30 d, and the set voltage varied little after 30 d (Fig. 6.6a).

However, the device without ZnO degraded within 3 d, so the on/off ratio cannot be maintained constantly, and the set voltage varied widely between 0.6 V and 1.8 V. This fast degradation may be a result of decomposition of $CH_3NH_3PbI_3$ to CH_3NH_3I and PbI_2. To achieve CMOS compatibility, a memory device was fabricated with a protective AlO_x layer that was deposited by atomic layer deposition. The device structure with $Al/AlO_x/CH_3NH_3PbI_3/ITO$ also maintained the bipolar resistive switching behavior after 30 days in air (Fig. 6.6b). These results indicate that an encapsulating layer deposited on perovskite helps to make it compatible with practical memory applications.

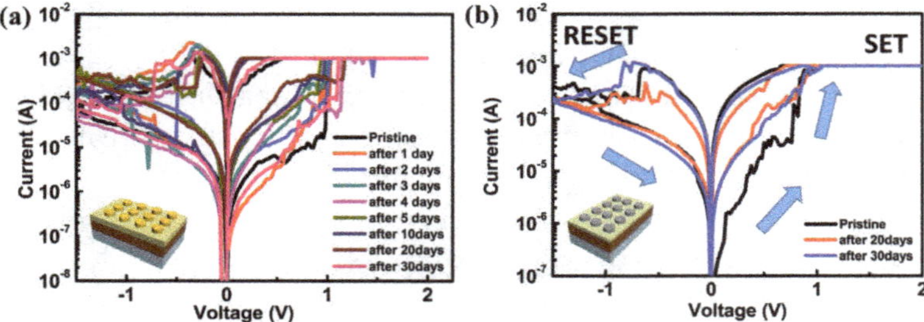

Figure 6.6 Stable resistive switching behavior of (a) ZnO-capped-perovskite memory devices and (b) AlO_x-capped perovskite memory device irrespective of storage time in ambient atmosphere. Reproduced with permission from Ref. [21], Copyright © 2017 Nature Publishing Group.

Inorganic halide perovskites, such as $CsPbBr_3$, can have better stability than OIPs. $CsPbBr_3$ has been applied as resistive switching layers. $CsPbBr_3$-based ReRAM has been demonstrated in an MIM structure of $Ni/CsPbBr_3/FTO$ with a capping ZnO layer [22]. The resistive switching effect of the ZnO capping device was derived from the I–V curve of the ReRAM; by negatively applying the voltage, the device was set at around −0.95 V, and by reversing the applied voltage, it was reset at ~0.71 V (Fig. 6.7a). The fabricated device showed a high on/off current ratio of $\sim10^5$ for 10^4 s, and the large on/off ratio was maintained for more than 20 days (Fig. 6.7b). These results show that a protective amorphous ZnO layer increases the environmental stability and data retention of devices. To examine the governing conduction mechanism of the ZnO-capped CsPbBr3-based ReRAM, the HRS followed three conduction mechanism: ohmic conduction, Schottky emission, and SCLC.

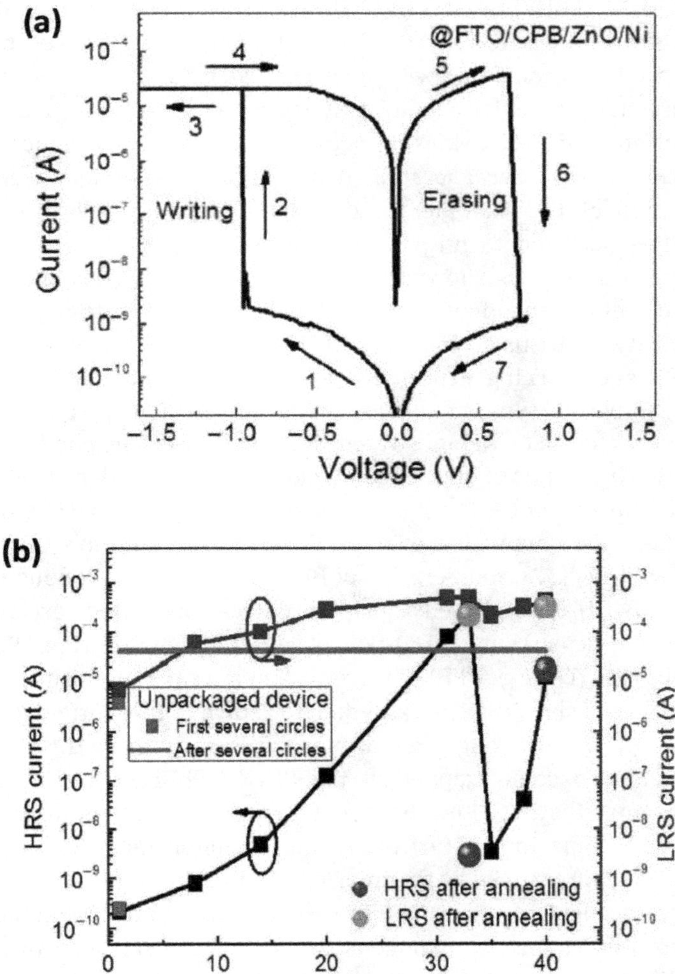

Figure 6.7 (a) *I–V* characteristics of resistive switching memory with FTO/ $CsPbBr_3$/ZnO/Ni structure. (b) Change of on/off ratio of the device in ambient air. Reprinted from Ref. [22], Copyright 2017, with permission of Springer Nature.

6.4.5 OIP-Based Neuromorphic Applications

The applications of OIP materials have been extended to neuromorphic devices. A two-terminal $CH_3NH_3PbI_3$ device with a

Au/$CH_3NH_3PbI_3$/PEDOT: PSS/ITO structure has been demonstrated in a synaptic device (Fig. 6.8a) [25]. Ion migration induced by an electric field yielded a switchable p-i-n structure; as a result, the conductivity/resistivity of this synaptic device can be changed. Application of positive bias (or negative bias) to the device increases its conductivity because applied biases function as forward bias (Fig. 6.8b,c). For example, the device flips to p-i-n polarity under positive bias and to n-i-p polarity under negative bias. The OIP device in this work showed various synaptic properties, including spike-timing-dependent plasticity (STDP), spike-rate-dependent plasticity, short-term plasticity (STP) and long-term potentiation (LTP), and learning experience behavior. In the other research, an OIP-based synaptic device with an Al/$CH_3NH_3PbBr_3$/BCCP structure has been demonstrated (Fig. 6.8d) [24]. The conductance of $CH_3NH_3PbBr_3$ can be changed by ion migration under an electric field, and the BCCP layer stores trapped mobile ions (Fig. 6.8e). Br^- has lower activation energy (E_A ~0.2 eV) than does $CH_3NH_3^+$ (E_A ~0.8 eV) [52], so migration of Br^- may change the conductivity. The $CH_3NH_3PbBr_3$-based synaptic device presented excitatory postsynaptic current, paired-pulse facilitation (PPF), STP, LTP, and STDP. The STP (e.g., PPF) occurred when a weak pulse caused ions to migrate a short distance and then to return to their original sites. Application of a strong pulse or many pulses can drive the ions far enough to become trapped at the $CH_3NH_3PbBr_3$/BCCP interface or to enter the BCCP and become trapped there. When the pulses stopped, some ions returned to their equilibrium positions, but some remain trapped at the interface and in the BCCP. As a result, many vacancy defects remain in $CH_3NH_3PbBr_3$ to form conductive paths; this change is analogous to LTP (Fig. 6.8f). Also the Ag/$CH_3NH_3PbI_{3-x}Cl_x$/FTO device underwent gradual potentiation and depression behavior that could be applied in neuromorphic devices [26]. These results show that applications of OIP materials can be extended to neuromorphic electronics.

6.4.6 Ion Distribution and Resistive Switching Effect under Electric Field and Light Illumination

Recently, OIP-based materials have emerged as effective switching layers that exploit ion migration inside the OIP films. However, for

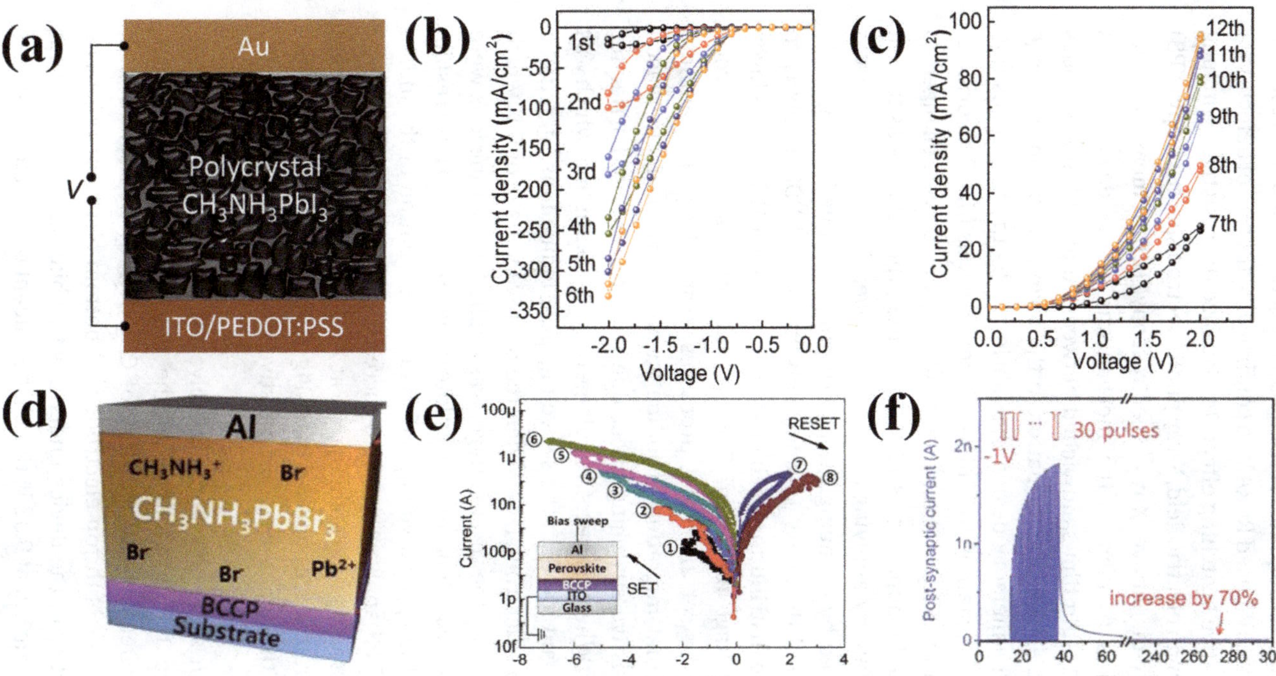

Figure 6.8 Memristive behavior of the OIP synaptic device. (a) Schematics of the device structure. (b, c) Memristive properties of the device under scanning, with positive and negative biases scanning at 0.1 Vs^{-1}. (d) Geometry of an OIP perovskite artificial synapse. (e) DC *I–V* sweeps with consecutively increased ranges that induce gradual setting and resetting processes. Inset: Schematic of the structure and measurement of an OIP-based artificial synapse. (f) Long-term potentiation achieved by application of 30 consecutive pulses to the artificial synapse. Reprinted from Refs. [24, 25], Copyright 2017, with permission from John Wiley & Sons.

elaborate memory applications, further research should investigate operation of OIP-based ReRAM to confirm distribution of defects that induce resistive switching effect. To visualize the resistive switching effect under an electric field, X-ray photoelectron spectroscopy (XPS) and energy-dispersive X-ray spectroscopy (EDS) measurements were applied to identify the mechanisms of the device [28]. A $CH_3NH_3PbI_3$ thin film sandwiched between Al and P^+-Si electrodes could store electrical information in the form of resistance-change memory. The memory device was fabricated by the fast-deposition crystallization method to achieve a homogeneous film. The original off state of the $CH_3NH_3PbI_3$-based memory device was changed to the on state by applying −3.15 V. Thereafter, the device was turned to its off state by applying a positive voltage; the set-reset process was repeated by changing the polarity of the applied voltage with the optimized thickness 140 nm. This study used EDS mapping to observe the distribution of the iodine ions in the $CH_3NH_3PbI_3$ layer, applying voltage of −5 and 5 V sequentially to a P^+-Si electrode to induce the set and reset processes, respectively. The authors suggested that migration of iodine vacancy was the main reason for the resistive switching effect due to a low energy barrier for migrations. When the negative voltage of −5 V was applied, which was smaller magnitude than the set voltage, the iodine ions aggregated near the top electrode (Al). When the positive voltage of 5 V, which was higher than the reset voltage, was applied, the iodine ions were distributed throughout the perovskite films (Fig. 6.9a). These results show that the applied voltage affects the distribution of the iodine ions. Also XPS analysis detected increasing iodine binding energy with an increase in the etching time at $V = 5$ V; these results indicate that I^- migrate under applied bias (Fig. 6.9b). These measurements supported the hypothesis that resistive switching was a result of I^- migration.

A recent study attempted to solve the chemical nature of the defects and the distribution of the defects that yield resistive switching effects. A 200 nm film of $CH_3NH_3PbI_3$ was sandwiched between Au as both top electrode and bottom electrode [27]. The I–V properties of the device under positive bias showed an abrupt increase in current at 0.32 V; this change meant that a stable LRS was

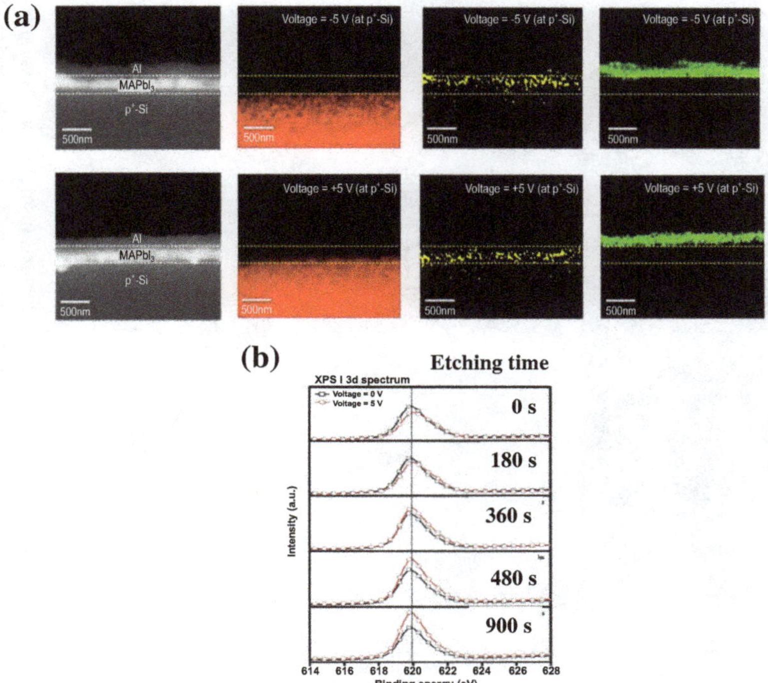

Figure 6.9 Optical and chemical analyses of a resistive switching device based on $CH_3NH_3PbI_3$ according to operation mechanism. SEM image and EDS mapping of Si, I, and Al atoms of the $CH_3NH_3PbI_3$-memory device sequential applied with (a) −5 V and 5 V, (b) XPS analysis of I 3d spectrum according to etching times for $CH_3NH_3PbI_3$ film with 0 V and 5 V. Reprinted from Ref. [28], Copyright 2017, with permission from John Wiley & Sons.

formed. The ReRAM based on $CH_3NH_3PbI_3$ maintained the LRS until a reverse bias of −0.13 V forced a reset. To identify the dominant ion process during resistive switching behavior, temperature-dependent retention measurements were presented with a $Au/CH_3NH_3PbI_3/Au$ structure at different temperatures (Fig. 6.10a). As the temperature decreased from 300 to 200 K, $\ln(t)$ versus $1/k_BT$ followed the linear relationship, which led to activation energy E_A of 0.17 eV, where (t) is the retention time, k_B is the Boltzmann constant, and T is the absolute temperature. This obtained value was consistent with the previous reports for V_I or I_i (0.1–0.6 eV), and this suggests that I^-

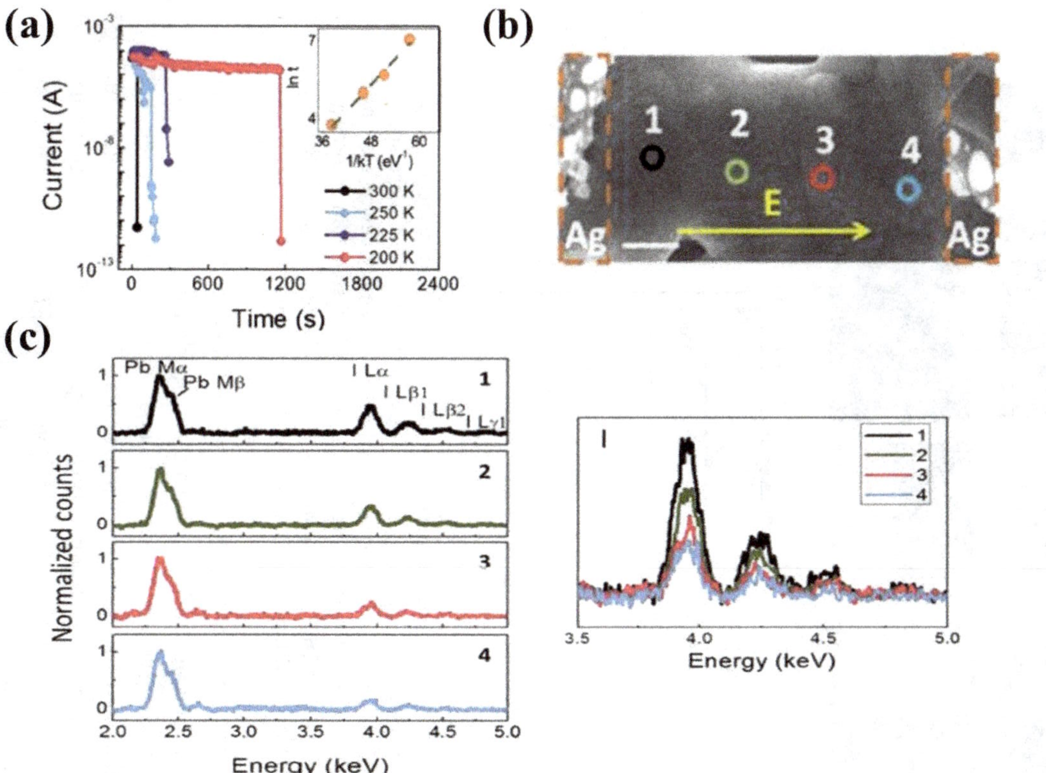

Figure 6.10 (a) Temperature-dependent LRS retention behavior of the metal/$CH_3NH_3PbI_3$/Au device structure. Inset: ln(t) vs. $1/k_BT$ plot, (b) SEM image of the device. Locations 1–4 mark the positions where the EDX analyses were held. (c) Left: EDX spectra of the device at an LRS, showing the main characteristic Pb M series peaks. Right: Comparison of the I peaks at locations 1–4. Reprinted from Ref. [27], Copyright 2017, with permission of John Wiley & Sons.

ion migration is a dominant process. The effect of the composition of the top electrode was investigated by replacing the Au top electrode with Ag. This change extended the LRS retention from 38 to 706 s and decreased V_{set}. To confirm the main ion source that induces resistive switching and observe the ion distribution, energy-dispersive X-ray spectroscopy (EDX) measurements were performed. In these measurements, a Ag/$CH_3NH_3PbI_3$/Ag planar structure was used to isolate the perovskite film between the electrodes. A voltage sweep was applied (the electric field direction was toward the right Ag electrode with compliance current of 10^{-4} A), the device switched to low resistance state, and the distributions of I:Pb at 1-4 positions along the film were measured. While resistive switching effect occurred, quantitative analysis of the I:Pb peak ratio decreased gradually (Fig. 6.10b,c). This phenomenon is related to the migration of I^- ions under the electric field, which results in resistive switching behavior. Retention measurements and previous theoretical calculations suggested that the I^- ion was the dominant ion to migrate. Also in the previous reports, the defect distribution in $CH_3NH_3PbI_3$ films was observed by light illumination [56]. The Ag/$CH_3NH_3PbI_{3-x}Cl_x$/FTO was exposed to 1 sun AM 1.5 simulated sunlight (100 mW·cm^{-2}); as a result, set and reset voltages decreased (Fig. 6.11a) [26]. Generation of excessive charge carriers by exposure to light might have induced this effect. When a direct-bandgap material $CH_3NH_3PbI_{3-x}Cl_x$ (~1.55 eV) was exposed to light, electron-hole pairs are generated. When voltage was applied to the device, the photogenerated electrons and injected electrons quickly occupied the trap sites. Thus, the trap-filled limit current law ($I \propto V^2$) became quickly applicable and ohmic conduction did not occur at low voltage (Fig. 6.11b). These results indicate that the low operating voltage is attributed to photogenerated electrons, which lead to fast formation of conductive paths. Another study measured the resistive switching effect after exposing white light–emitting diodes to different intensities of light [28]. When the device was exposed to lights, the current increased but the LRS did not change (Fig. 6.11c). At the highest light intensities (310 lux), the HRS and the LRS became the same due to increased flow of electrons after illumination. Light illumination affects ion migration [27]. To study the effect of light illumination, the Ag/$CH_3NH_3PbI_3$/Au planar device structure was used. As the illumination intensity was increased from 0 to 1.29 μW·cm^{-2}, the set voltage was increased from 2.1 to 3.8 V. Moreover, the set voltage was increased and LRS retention failure happened fast under illumination with high-intensity laser (4–12 mW·cm^{-2}, 525 nm). These results suggest that illumination prevents the perovskite film from forming a V_I-rich conductive filament; this conclusion is consistent with previous studies of PL [56, 57]. Under illumination, the V_I^+ in $CH_3NH_3PbI_3$ is unstable and spontaneously joins with I_i^-, which inhibits the formation of V_I-rich conductive filaments. These phenomena were exploited in an electrical-write and optical-erase memory device. The set process was induced by an electrical bias, and the reset process was induced by illumination.

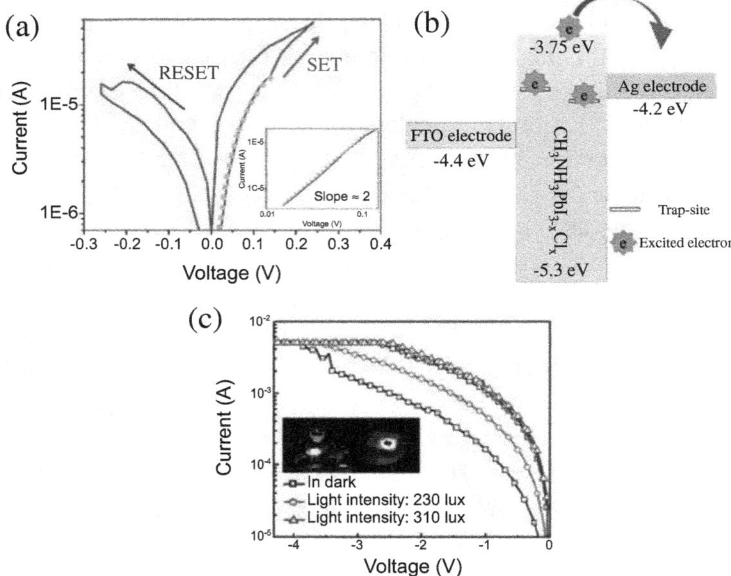

Figure 6.11 (a) Log (*I*)–(*V*) properties of both positive- and negative-bias regions under light illumination. Inset: Log (*I*)–log (*V*) plot of a high-resistance state in the range from 0 V to 0.15 V. (b) Energy diagram of the Ag/$CH_3NH_3PbI_{3-x}Cl_x$/FTO device. Reproduced from Ref. [26], Copyright 2016, with permission of The Royal Society of Chemistry. Effect of light on the $CH_3NH_3PbI_3$ memory device (ITO/$CH_3NH_3PbI_3$/p$^+$-Si); (c) *I*–*V* properties of the device measured under dark and illuminated conditions with an intensity of 230 and 310 lux; V_I dynamics effected by light illumination. Reprinted from Ref. [28], Copyright 2017, with permission of John Wiley & Sons.

6.4.7 Two-Dimensional OIP-Based ReRAM

The most studied OIPs for memory application have been (3D) $CH_3NH_3PbI_3$, $CH_3NH_3PbI_{3-x}Cl_x$, $CH_3NH_3PbBr_{3-x}$, and $CsPbBr_3$. These materials have multilevel resistive switching behavior and also a high on/off ratio (~10^6). However, 3D OIP-based memory devices still lack reliability and stability compared to resistive memories based on inorganic materials. A recent study introduced a 2D perovskite material as a resistive switching layer [29]. The layered 2D perovskite form has a general formula of $(RNH_3)_2(CH_3NH_3)_{m-1}A_mX_{3m+1}$, where R is an alkyl or aromatic moiety, A is a metal cation, and X is a halide anion. The 2D perovskites of $(C_4H_9NH_3)_2PbI_4$ (BA_2PbI_4) are composed of inorganic layers of $[PbI_6]^{2-}$ octahedra sandwiched between bilayers of intercalated organic chains such as butyl ammonium cations (Fig. 6.12a) [58, 59]. This study observed the change of the resistive switching effect as decreasing the dimension from 3D ($CH_3NH_3PbI_3$) to 2D (BA_2PbI_4). The ReRAM based on the 2D OIP showed bipolar resistive switching behavior.

The set voltage was represented by an F_E, and the on/off ratio was compared between 3D and 2D ReRAM. BA_2PbI_4 switched the device at the lowest set F_E (0.25×10^6 V/m) (Fig. 6.12b). As the dimensionality decreased from 3D to 2D, the on/off ratio increased (Fig. 6.12c). The anisotropic 2D structure achieved stable endurance without failure for 250 cycles compared to the 3D structure. Compared to the 3D structure, higher Schottky barrier heights and E_A led to smaller HRS current and a higher on/off ratio. Solution-processed BA_2PbI_4 was demonstrated on a 4-inch wafer and achieved reliable memory property. These results show that 2D perovskite could be a promising material for memory applications.

6.5 Summary and Outlook

OIP materials present exceptional optical and electronic properties, particularly the observation that defect migration causes hysteresis in the current-voltage curve. Moreover, by utilizing the defects in the perovskite materials, research has revealed various memory properties, such as multilevel and analog switching. In this chapter, we have summarized recent reports about OIP materials, especially the memory applications like flexible ReRAM, multilevel ReRAM, air-stable ReRAM encapsulated with metal-oxide layers, and neuromorphic applications. Theoretical results have identified iodine (halide) vacancies as the dominant mobile ionic defect that is related to hysteresis, and these vacancies are the major reason for resistive switching.

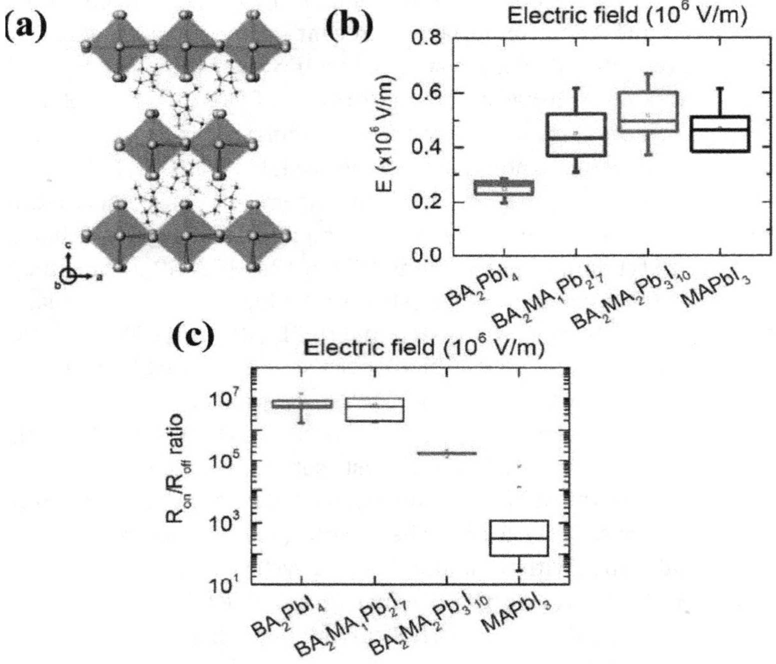

Figure 6.12 (a) Schematic image of 2D BA_2PbI_4, (b) switching voltage, represented by electric field (F_E), for the set process for BA_2PbI_4, $BA_2MAPb_2I_7$, $BA_2MA_2Pb_3I_{10}$, and $MAPbI_3$, and (c) the on/off ratio for each material (reading voltage 0.02 V). Reproduced from Ref. [31], Copyright 2017, with permission of The Royal Society of Chemistry.

Before commercialization of OIP-based ReRAM is viable, problems such as poor stability, toxicity issues, and memory reliability must be solved. Long-term stability is the main deficit compared with stable inorganic materials, such as silicon. Although an OIP-based ReRAM with a metal-oxide layer was presented in this chapter, device stability should be further increased before practical applications are feasible. Moreover, lead-free OIPs with good memory properties should be developed. However, the superiority of lead has resulted in better memory characteristics of OIP-based ReRAM, so this replacement with lead-free OIPs remains difficult. The reason for the better memory properties of lead-based OIP applications requires further study.

Although various OIP-based ReRAM devices have been demonstrated, their memory storage capabilities must be improved before they can be practical information-storage devices. To increase the potential of OIP-based ReRAM for future electronic applications, methods should be found to control the type and concentrations of ions and their pathways through the perovskite films. Also, memory responses, such as fast switching, long retention, and a large on/off ratio, must be optimized to obtain reliable and predictable memory devices. If these scientific and technical problems can be solved, we believe that OIP materials will facilitate advances in information-storage applications.

References

1. Lee, M. M., Teuscher, J., Miyasaka, T., Murakami, T. N., and Snaith, H. J. (2012). Efficient hybrid solar cells based on meso-superstructured organometal halide perovskites. *Science*, **338**, 643–647.
2. Burschka, J., Pellet, N., Moon, S. J., Humphry-Baker, R., Gao, P., et al. (2013). Sequential deposition as a route to high-performance perovskite-sensitized solar cells. *Nature*, **499**, 316.
3. Zuo, C., Bolink, H. J., Han, H., Huang, J., Cahen, D., and Ding, L. (2016). Advances in perovskite solar cells. *Adv. Sci.*, **3**, 1500324.
4. Wehrenfennig, C., Eperon, G. E., Johnston, M. B., Snaith, H. J., and Herz, L. M. (2014). High charge carrier mobilities and lifetimes in organolead trihalide perovskites. *Adv. Mater.*, **26**, 1584–1589.
5. Xing, G. C., Mathews, N., Sun, S. Y., Lim, S. S., Lam, Y. M., et al. (2013). Long-range balanced electron- and hole-transport lengths in organic-inorganic $CH_3NH_3PbI_3$. *Science*, **342**, 344–347.
6. Kim, Y. H., Cho, H., Heo, J. H., Kim, T. S., Myoung, N., et al. (2015). Multicolored organic/inorganic hybrid perovskite light-emitting diodes. *Adv. Mater.*, **27**, 1248–1254.

7. Dou, L. T., Yang, Y., You, J. B., Hong, Z. R., Chang, W. H., et al. (2014). Solution-processed hybrid perovskite photodetectors with high detectivity. *Nat. Commun.*, **5**, 5404.

8. Chen, S., Teng, C. J., Zhang, M., Li, Y. R., Xie, D., and Shi, G. Q. (2016). A flexible UV-Vis-NIR photodetector based on a perovskite/conjugated-polymer composite. *Adv. Mater.*, **28**, 5969.

9. Xiao, Z. G., Yuan, Y. B., Shao, Y. C., Wang, Q., Dong, Q. F., et al. (2015). Giant switchable photovoltaic effect in organometal trihalide perovskite devices. *Nat. Mater.*, **14**, 193–198.

10. Waser, R., and Aono, M. (2007). Nanoionics-based resistive switching memories. *Nat. Mater.*, **6**, 833–840.

11. Linn, E., Rosezin, R., Kugeler, C., and Waser, R. (2010). Complementary resistive switches for passive nanocrossbar memories. *Nat. Mater.*, **9**, 403–406.

12. Waser, R., Dittmann, R., Staikov, G., and Szot, K. (2009). Redox-based resistive switching memories - nanoionic mechanisms, prospects, and challenges. *Adv. Mater.*, **21**, 2632.

13. Gao, S., Song, C., Chen, C., Zeng, F., and Pan, F. (2012). Dynamic processes of resistive switching in metallic filament-based organic memory devices. *J. Phys. Chem. C*, **116**, 17955–17959.

14. Yang, J. J., Pickett, M. D., Li, X. M., Ohlberg, D. A. A., Stewart, D. R., and Williams, R. S. (2008). Memristive switching mechanism for metal/oxide/metal nanodevices. *Nat. Nanotechnol.*, **3**, 429–433.

15. Yasuhara, R., Yamamoto, T., Ohkubo, I., Kumigashira, H., and Oshima, M. (2010). Interfacial chemical states of resistance-switching metal/$Pr_{0.7}Ca_{0.3}MnO_3$ interfaces. *Appl. Phys. Lett.*, **97**, 132111.

16. Yoo, E. J., Lyu, M., Yun, J. H., Kang, C. J., Choi, Y. J., and Wang, L. (2015). Resistive switching behavior in organic-inorganic hybrid $CH_3NH_3PbI_{3-x}Cl_x$ perovskite for resistive random access memory devices. *Adv. Mater.*, **27**, 6170–6175.

17. Gu, C., and Lee, J.-S. (2016). Flexible hybrid organic-inorganic perovskite memory. *ACS Nano*, **10**, 5413–5418.

18. Liu, D. J., Lin, Q. Q., Zang, Z. G., Wang, M., Wangyang, P. H., et al. (2017). Flexible all-inorganic perovskite $CsPbBr_3$ nonvolatile memory device. *ACS Appl. Mater. Interfaces*, **9**, 6171–6176.

19. Yang, K. Y., Li, F. S., Veeramalai, C. P., and Guo, T. L. (2017). A facile synthesis of $CH_3NH_3PbBr_3$ perovskite quantum dots and their application in flexible nonvolatile memory. *Appl. Phys. Lett.*, **110**, 083102.

20. Hwang, B. H., Gu, C. W., Lee, D. H., and Lee, J.-S. (2017). Effect of halide-mixing on the switching behaviors of organic-inorganic hybrid perovskite memory. *Sci. Rep.*, **7**, 43794.

21. Hwang, B., and Lee, J.-S. (2017). Hybrid organic-inorganic perovskite memory with long-term stability in air. *Sci. Rep.*, **7**, 673.

22. Wu, Y., Wei, Y., Huang, Y., Cao, F., Yu, D. J., et al. (2017). Capping $CsPbBr_3$ with ZnO to improve performance and stability of perovskite memristors. *Nano Res.*, **10**, 1584–1594.

23. Hwang, B., and Lee, J.-S. (2017). A strategy to design high-density nanoscale devices utilizing vapor deposition of metal halide perovskite materials. *Adv. Mater.*, **29**, 170104.

24. Xu, W., Cho, H., Kim, Y. H., Kim, Y. T., Wolf, C., et al. (2016). Organometal halide perovskite artificial synapses. *Adv. Mater.*, **28**, 5916.

25. Xiao, Z. G., and Huang, J. S. (2016). Energy-efficient hybrid perovskite memristors and synaptic devices. *Adv. Electron. Mater.*, **2**, 1600100.

26. Yoo, E., Lyu, M., Yun, J. H., Kang, C., Choi, Y., and Wang, L. Z. (2016). Bifunctional resistive switching behavior in an organolead halide perovskite based Ag/$CH_3NH_3PbI_{3-x}Cl_x$/FTO structure. *J. Mater. Chem. C*, **4**, 7824–7830.

27. Zhu, X. J., Lee, J., and Lu, W. D. (2017). Iodine vacancy redistribution in organic-inorganic halide perovskite films and resistive switching effects. *Adv. Mater.*, **29**, 1700527.

28. Kim, D. J., Tak, Y. J., Kim, W. G., Kim, J. K., Kim, J. H., and Kim, H. J. (2017). Resistive switching properties through iodine migrations of a hybrid perovskite insulating layer. *Adv. Mater. Interfaces*, **4**, 1601035.

29. Seo, J.-Y., Choi, J., Kim, H.-S., Kim, J., Yang, J.-M., et al. (2017). Wafer-scale reliable switching memory based on 2-dimensional layered organic-inorganic halide perovskite. *Nanoscale*, **9**, 15278–15285.

30. Yan, K., Chen, B. X., Hu, H. W., Chen, S., Dong, B., et al. (2016). First fiber-shaped non-volatile memory device based on hybrid organic-inorganic perovskite. *Adv. Electron. Mater.*, **2**, 1600160.

31. Muthu, C., Agarwal, S., Vijayan, A., Hazra, P., Jinesh, K. B., and Nair, V. C. (2016). Hybrid perovskite nanoparticles for high-performance resistive random access memory devices: control of operational parameters through chloride doping. *Adv. Mater. Interfaces*, **3**, 1600092.

32. Xu, Z. Q., Liu, Z. H., Huang, Y., Zheng, G. H. J., Chen, Q., and Zhou, H. P. (2017). To probe the performance of perovskite memory devices: defects property and hysteresis. *J. Mater. Chem. C*, **5**, 5810–5817.

33. Ercan, E., Chen, J.-Y., Tsai, P.-C., Lam, J.-Y., Huang, S. C.-W., et al. (2017). A redox-based resistive switching memory device consisting of organic–inorganic hybrid perovskite/polymer composite thin film. *Adv. Electron. Mater.*, **3**, 1700344.

34. Choi, J., Park, S., Lee, J., Hong, K., Kim, D. H., et al. (2016). Organolead halide perovskites for low operating voltage multilevel resistive switching. *Adv. Mater.*, **28**, 6562.

35. Akinaga, H., and Shima, H. (2010). Resistive random access memory (ReRAM) based on metal oxides. *Proc. IEEE*, **98**, 2237–2251.

36. Pan, F., Chen, C., Wang, Z. S., Yang, Y. C., Yang, J., and Zeng, F. (2010). Nonvolatile resistive switching memories-characteristics, mechanisms and challenges. *Prog. Nat. Sci.*, **20**, 1–15.

37. Kalaev, D., Yalon, E., and Riess, I. (2015). On the direction of the conductive filament growth in valence change memory devices during electroforming. *Solid State Ionics*, **276**, 9–17.

38. Sanchez, R. S., Gonzalez-Pedro, V., Lee, J. W., Park, N. G., Kang, Y. S., et al. (2014). Slow dynamic processes in lead halide perovskite solar cells. Characteristic times and hysteresis. *J. Phys. Chem. Lett.*, **5**, 2357–2363.

39. Miller, D. W., Eperon, G. E., Roe, E. T., Warren, C. W., Snaith, H. J., and Lonergan, M. C. (2016). Defect states in perovskite solar cells associated with hysteresis and performance. *Appl. Phys. Lett.*, **109**, 153902.

40. Tress, W., Marinova, N., Moehl, T., Zakeeruddin, S. M., Nazeeruddin, M. K., and Gratzel, M. (2015). Understanding the rate-dependent J-V hysteresis, slow time component, and aging in $CH_3NH_3PbI_3$ perovskite solar cells: the role of a compensated electric field. *Energy Environ. Sci.*, **8**, 995–1004.

41. Frost, J. M., and Walsh, A. (2016). What is moving in hybrid halide perovskite solar cells?. *Acc. Chem. Res.*, **49**, 528–535.

42. Elumalai, N. K., and Uddin, A. (2016). Hysteresis in organic-inorganic hybrid perovskite solar cells. *Sol. Energy Mater. Sol. Cells*, **157**, 476–509.

43. Yang, T. Y., Gregori, G., Pellet, N., Gratzel, M., and Maier, J. (2015). The significance of ion conduction in a hybrid organic-inorganic lead-iodide-based perovskite photosensitizer. *Angew. Chem. Int. Ed.*, **54**, 7905–7910.

44. Yun, J. S., Seidel, J., Kim, J., Soufiani, A. M., Huang, S. J., et al. (2016). Critical role of grain boundaries for ion migration in formamidinium and methylammonium lead halide perovskite solar cells. *Adv. Energy Mater.*, **6**, 1600330.

45. Xing, J., Wang, Q., Dong, Q. F., Yuan, Y. B., Fanga, Y. J., and Huang, J. S. (2016). Ultrafast ion migration in hybrid perovskite polycrystalline thin films under light and suppression in single crystals. *Phys. Chem. Chem. Phys.*, **18**, 30484–30490.

46. Yu, H., Lu, H. P., Xie, F. Y., Zhou, S., and Zhao, N. (2016). Native defect-induced hysteresis behavior in organolead iodide perovskite solar cells. *Adv. Funct. Mater.*, **26**, 1411–1419.

47. Li, C., Tscheuschner, S., Paulus, F., Hopkinson, P. E., Kiessling, J., et al. (2016). Iodine migration and its effect on hysteresis in perovskite solar cells. *Adv. Mater.*, **28**, 2446–2454.

48. Snaith, H. J., Abate, A., Ball, J. M., Eperon, G. E., Leijtens, T., et al. (2014). Anomalous hysteresis in perovskite solar cells. *J. Phys. Chem. Lett.*, **5**, 1511–1515.

49. Shao, Y. H., Xiao, Z. G., Bi, C., Yuan, Y. B., and Huang, J. S. (2014). Origin and elimination of photocurrent hysteresis by fullerene passivation in $CH_3NH_3PbI_3$ planar heterojunction solar cells. *Nat. Commun.*, **5**, 5784.

50. Kutes, Y., Ye, L. H., Zhou, Y. Y., Pang, S. P., Huey, B. D., and Padture, N. P. (2014). Direct observation of ferroelectric domains in solution-processed $CH_3NH_3PbI_3$ perovskite thin films. *J. Phys. Chem. Lett.*, **5**, 3335–3339.

51. Unger, E. L., Hoke, E. T., Bailie, C. D., Nguyen, W. H., Bowring, A. R., et al. (2014). Hysteresis and transient behavior in current-voltage measurements of hybrid-perovskite absorber solar cells. *Energy Environ. Sci.*, **7**, 3690–3698.

52. Azpiroz, J. M., Mosconi, E., Bisquert, J., and De Angelis, F. (2015). Defect migration in methylammonium lead iodide and its role in perovskite solar cell operation. *Energy Environ. Sci.*, **8**, 2118–2127.

53. Li, C., Guerrero, A., Zhong, Y., and Huettner, S. (2017). Origins and mechanisms of hysteresis in organometal halide perovskites. *J. Phys. Condens. Matter*, **29**, 193001.

54. Eames, C., Frost, J. M., Barnes, P. R. F., O'Regan, B. C., Walsh, A., and Islam, M. S. (2015). Ionic transport in hybrid lead iodide perovskite solar cells. *Nat. Commun.*, **6**.

55. Yuan, Y. B., and Huang, J. S. (2016). Ion migration in organometal trihalide perovskite and its impact on photovoltaic efficiency and stability. *Acc. Chem. Res.*, **49**, 286–293.

56. Dequilettes, D. W., Zhang, W., Burlakov, V. M., Graham, D. J., Leijtens, T., et al. (2016). Photo-induced halide redistribution in organic-inorganic perovskite films. *Nat. Commun.*, **7**, 7497.

57. Mosconi, E., Meggiolaro, D., Snaith, H. J., Stranks, S. D., and De Angelis, F. (2016). Light-induced annihilation of Frenkel defects in organo-lead halide perovskites. *Energy Environ. Sci.*, **9**, 3180–3187.

58. Dou, L. T., Wong, A. B., Yu, Y., Lai, M. L., Kornienko, N., et al. (2015). Atomically thin two-dimensional organic-inorganic hybrid perovskites. *Science*, **349**, 1518–1521.

59. Cao, D. H., Stoumpos, C. C., Farha, O. K., Hupp, J. T., and Kanatzidis, M. G. (2015). 2D homologous perovskites as light-absorbing materials for solar cell applications. *J. Am. Chem. Soc.*, **137**, 7843–7850.

Chapter 7

Carbon-Based Large-Scale Technology

Hongwei Han, Yaoguang Rong, Yue Hu, Anyi Mei,
Yuli Xiong, and Chengbo Tian
Wuhan National Laboratory for Optoelectronics,
Huazhong University of Science and Technology,
1037 Luoyu Road, Wuhan, Hubei, China
hongwei.han@mail.hust.edu.cn

7.1 Introduction

Hybrid organic–inorganic perovskite solar cells (PSCs) with practical levels of power conversion efficiency (PCE) first emerged in 2009 through the pioneering work of Miyasaka [1] using mesoporous TiO_2 film combined with $CH_3NH_3PbI_3$ perovskite, simulating a boosted growth in the academic study of these devices. With a certified high-power conversion efficiency (PCE) over 22% [2], these devices are recognized as serious contenders to rival the leading photovoltaic technologies [3–5]. The configuration of PSCs merely changed since the first solid-state PSC was introduced in 2012 [6, 7]. The basic PSC has several main components: (1) a conductive

Multifunctional Organic–Inorganic Halide Perovskite: Applications in Solar Cells, Light-Emitting Diodes, and Resistive Memory
Edited by Nam-Gyu Park and Hiroshi Segawa

ISBN 978-981-4800-52-5 (Hardcover), 978-1-003-27593-0 (eBook)
www.jennystanford.com

transparent glass, typically fluorine-doped tin dioxide (SnO_2:F, FTO) or indium-doped tin oxide (SnO_2:In, ITO); (2) an electron-selective contact; [8] (3) the perovskite absorbing layer; (4) a hole-selective contact; and (5) a metal contact, typically Au [9]. Depending on the n-i-p configuration [10, 11] or p-i-n configuration [12] of PSC, the sequence of depositing the electron-selective contact and the hole-selective contact can be switched. Ongoing studies addressed some of the key challenges in this field, including further cost reduction [13], stability improvements [14, 15], scale-up [16] and manufacture [17], environmental friendliness [18–20] and a deep understanding of the thermodynamic [21, 22] and kinetic mechanism [23].

Among all, carbon-based large-scale technology based on printing technique has attracted attention both in scientific and business society. Carbon materials, particularly graphite and carbon black, possess advantages of accessibility, low cost, high electrical conductivity, chemical stability, water resistance and environmental friendliness [24, 25]. This chapter aims to give an overview of these advances by systematically summarizing recent research progresses and the current state of commercialization activities.

7.2 Device Structures and Working Principles

The general working principles and energy level diagram of carbon-based PSCs are illustrated in Fig. 7.1a. The energy alignment is important as it facilitates the injection of electrons and blocks the recombination of electrons with holes at the back contact. Upon light illumination, the electrons and holes are generated in the conduction band (CB) and the valence band (VB) of the perovskite. The built-in electric field in this depletion region directs from TiO_2 to perovskite, which is beneficial for the electron-hole separation and provides preferred directionality for the carriers. The photogenerated electrons are then injected into TiO_2 and quickly transfers to the conducting substrate, while the holes are extracted by the HTM or carbon back contact.

According to the device structure and the fabrication process, the carbon-based PSCs can be divided into three categories, as illustrated in Fig. 7.1b–d. It has been accepted by the scientific society that perovskite materials can transport holes effectively. Taking advantages of that, the expensive HTM can be omitted in PSCs. In addition, although solid-state PSCs show much enhanced stability than the PSCs using liquid electrolytes, the stability issue is still one of the biggest challenges for PSCs so far and the organic HTM is one of the weak component in the structure. Figure 7.1b illustrates the HTM-free PSCs using mesoscopic carbon. This structure was first developed by Han group in 2013 [26]. It comprises of an initially deposited triple-layer mesoporous scaffold, which is subsequently infiltrated by a perovskite precursor solution. The triple layers include a mesoscopic electron selecting layer (e.g. TiO_2, SnO_2, $BaSnO_3$, etc.), a mesoscopic carbon layer to collect the holes and a mesoscopic insulating layer (e.g. ZrO_2, Al_2O_3, etc.) that separates TiO_2 and carbon to prevent shunting. The mesopores are created by annealing the scaffold at a high temperature ≥400°C. Han group pioneered in this structure and explored both the mesoscopic scaffold and the perovskite absorber that fits the structure [26]. With a highest certified PCE of 12.8%, the PSCs based on this structure has been the most stable PSCs reported so far [27]. Studied by various groups around the world, the meso-carbon based PSCs have passed a number of stability tests. Namely, the continuous one-sun illumination test over 10000 h, the outdoor test in the hot desert (Jeddah, Saudi Arabia) over a week, the outdoor test in high-humidity environment (Wuhan, China) over a month [28], the thermal stability test at 100°C over 1500 h and the shelf-life stability test over one year [28].

Figure 7.1c shows the HTM-free PSCs using planar carbon electrode. The compact TiO_2 is usually deposited via spray pyrolysis. Then mesoporous TiO_2 layer with around 300 nm thickness is deposited by spin-coating and sintered. Perovskite layer can be prepared by various processing techniques, such as sequential deposition, single-step deposition, antisolvent methods, vapor-assisted methods and so on. Finally, the carbon electrode is deposited by direct spin coating of carbon materials such as carbon nanotube, graphene or carbon black. In other works, the carbon electrode is prepared on another substrate and subsequently transferred onto the device. The carbon electrode can be also processed by doctor-blading or screen-printing a paste or paint of carbon materials, which contains graphite, carbon black and some additives. Regardless of the carbon electrode types, they can be dried at low temperatures (≤100°C).

Hole-transport materials can be also added to the carbon-based PSCs. The conventional device structure of carbon-based PSCs is derived from the Au-based PSCs, as shown in Fig. 7.1d. It consists FTO glass/compact TiO_2 layer/mesoporous TiO_2 layer/perovskite/HTM/carbon. The highest PCE reported based on this structure is 17.0% by Sun group [29]. In that piece of work, they used the mixed-cations perovskite $(FAPbI_3)_{0.85}(MAPbBr_3)_{0.15}$ as the light absorber, a facile in-situ solid-state synthesized polymer poly(3,4-ethylenedioxythiophene) (PEDOT) as the HTM and a commercial carbon paste which was deposited by doctor-blading.

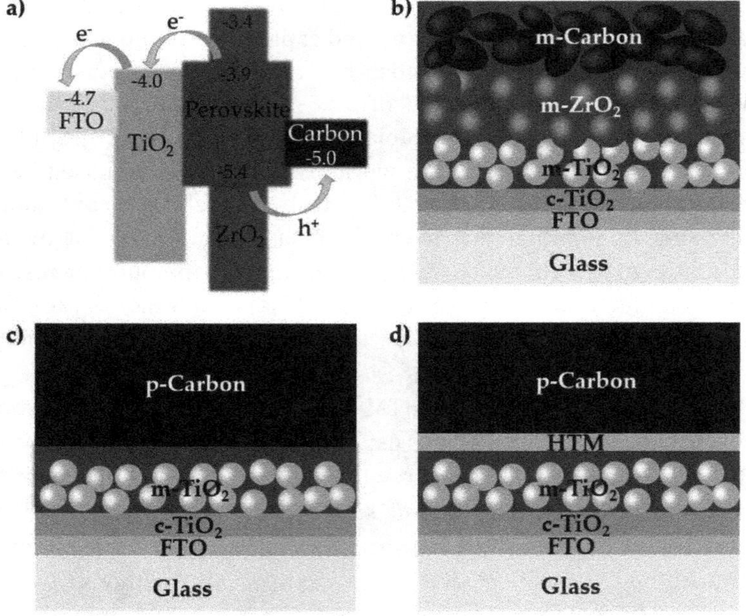

Figure 7.1 (a) Energy level and charge transfer behavior in carbon-based PSCs. (b) Carbon-based PSCs with structure FTO glass/compact TiO_2 layer/mesoporous TiO_2 layer/mesoporous ZrO_2 layer/perovskite/mesoporous carbon layer. (c) Carbon-based PSCs with structure FTO glass/compact TiO_2 layer/mesoporous TiO_2 layer/perovskite/carbon. (d) Carbon-based PSCs with structure FTO glass/compact TiO_2 layer/mesoporous TiO_2 layer/perovskite/HTM/carbon.

7.3 Pursuing High Efficiency and Stability

7.3.1 HTM-Free Mesoporous Carbon-Based PSCs

Perovskites in HTM-free MPSCs not only act as light absorber to generate electrons and holes upon light excitation, but also transport electrons and holes. Hence, a high-quality perovskite layer with large crystal size and less crystal boundary is of significance to achieve efficient charge transportation and reduced recombination for the high-efficiency HTM-free MPSCs.

As mentioned previously, Han's group firstly developed a fully printable HTM-free MPSCs based on mesoscopic triple-layer architecture in 2013. To improve the performance of the MPSCs, Mei et al. employed 5-ammoniumvaleric acid (5-AVA) into the conventional perovskite precursor solution, and developed a mixed-cation perovskite $(5\text{-AVA})_x(MA)_{1-x}PbI_3$ crystals with lower defect concentration and better pore filling as well as more complete contact with the TiO_2 scaffold and carbon counter electrode, yielding a longer exciton lifetime and a higher quantum yield for photoinduced charge separation as compared to $MAPbI_3$ [27]. Consequently, the MPSCs yielded a certified efficiency of 12.84%. More importantly, the device showed an excellent stability for over 1000 hours without encapsulation in ambient air under light soaking, as shown in Fig. 7.2a–c. Meanwhile, an extensive stability tests outdoor test for the device was conducted in a hot and humid climate (Jeddah, Saudi Arabia) and no degradation was observed. With the same device configuration, Ito group [30] proved the thermal stability of the MPSCs over 1500 h at 100°C. The large-area (10 × 10 cm^2) printable mesoscopic perovskite solar module yielded exceeding 10% efficiency, exhibiting local outdoor stability of 1 month and a shelf-life stability of over one year. Recently, a remarkable long-term stability of >10,000 h was achieved for the same cell [31]. In addition, the sustainable recycling of the PSCs in situ without components reconstruction or replacement is achieved by post-treating the photodegraded $MAPbI_3$ HTM-free PSCs via the methylamine gas, further enhancing the lifetime of PSCs [32]. Therefore, these promising results bode well for future practical applications of perovskite solar cells.

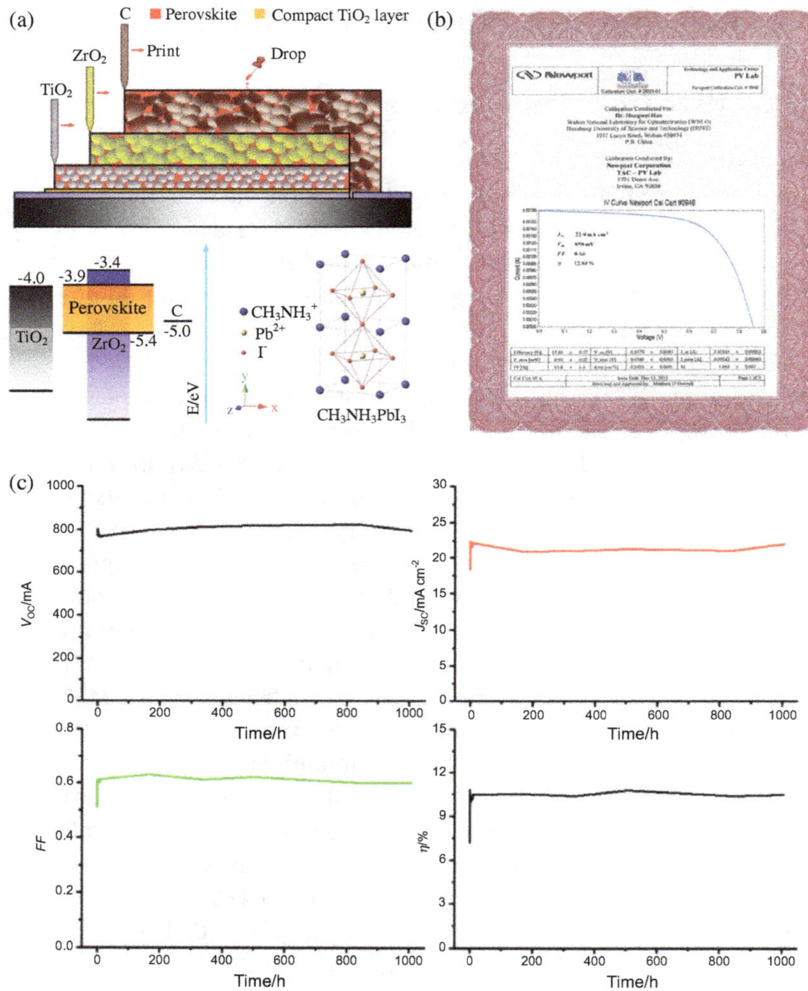

Figure 7.2 (a) Schematic drawing of the cross section and energy band diagram of the triple-layer MPSCs, and the crystal structure of $MAPbI_3$. (b) The accredited PV laboratory confirming a power conversion efficiency of 12.84% for the triple layer hole conductor-free mixed 5-AVA-MA perovskite solar cell. (c) Stability test of a triple-layer $(5\text{-AVA})_x(MA)_{(1-x)}PbI_3$ MPSCs under full AM 1.5 simulated sunlight in ambient air over 1008 h without encapsulation. Reproduced with permission from Ref. [27]. Copyright 2014, American Association for the Advancement of Science.

To further elevate the MPSCs' performance, Han et al. have successively conducted a series of works on improving the crystallinity and conductivity of the perovskite layer to enhance the PCE of the MPSCs. Hou et al. employed a multifunctional additive, guanidinium chloride, to improve the quality of the $CH_3NH_3PbI_3$ perovskite, and suppress the recombination of carriers in the device, which obtained a high efficiency of 14.27% with an obviously enhanced open-circuit voltage [33]. Meanwhile, Rong et al. systematically investigated synergy of ammonium chloride and moisture on perovskite crystallization for the MPSCs [34]. As shown in Fig. 7.3a–c, when the as-annealed devices were exposed to ambient air, the PCE increased dramatically in the first 10–48 h, and then remained stable (relative humidity (RH) 35%), decreased slightly (RH45% and RH55%), or decayed sharply (RH65%). The formation and transition of intermediate $CH_3NH_3X{\cdot}NH_4PbX_3(H_2O)_2$ (X = I or Cl) enabled high-quality perovskite $CH_3NH_3PbI_3$ crystals with preferential growth orientation. Consequently, the obtained MPSCs yielded a steady PCE of 15.45% and ambient air lifetime of over 130 days.

Additionally, Chen et al. studied influences of solvent on MPSCs with pristine $MAPbI_3$ via one-step deposition. It is found that the polarity and viscosity of solvents possessed a significant effect on the wettability of perovskite precursor solution and the stability of intermediate phases, which further influenced the crystallization and infiltration of perovskite in mesoporous films. Finally, due to the suitable interplay and compromise of polarity, viscosity, wettability and coordination ability, the DMF/DMSO solvent-based device achieved a promising PCE of 13.89% [35].

Other additives can improve the bulk conductivity of the perovskite and thus highly elevate the FF of the devices. Chen et al. developed a mixed-anion perovskite $MAPbI_{(3-x)}(BF_4)_x$ perovskite with an improved light harvesting ability, carrier concentration and conductivity [36]. The devices gain an enhanced photovoltaic performance with a dramatically improved fill factor (FF) of 0.76, due to more efficient charge transportation and suppressed recombination. Sheng et al. incorporated LiCl into the pristine $MAPbI_3$ perovskite precursor solution and demonstrated the effect of LiCl on the performance of MPSCs. The LiCl-mixed perovskite exhibited superior electronic properties by virtue of the elevated conductivity of the perovskite layer enabling faster carrier transport. LiCl-mixing also improved the crystallinity and morphology of the perovskite layer. Correspondingly, the LiCl-based MPSCs yielded an enhanced PCE of 14.5% and FF of 0.77 [37].

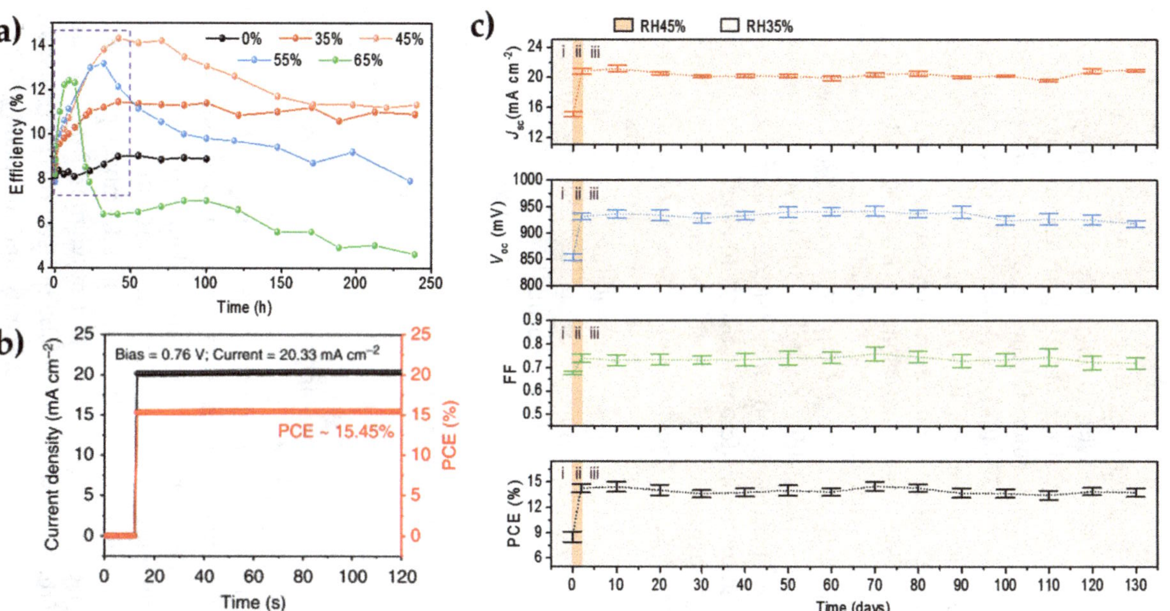

Figure 7.3 (a) PCE variations of as-annealed devices stored under ambient conditions with RH5–65%. The significant improvement in efficiency is marked with a purple rectangle. (b) Stabilized output of the device measured under a bias of 0.76 V, presenting a current density of 20.33 mA/cm^2 and a PCE of 15.45%. (c) Variations of J_{SC}, V_{OC}, FF, and PCE of devices upon fabricating and ageing under ambient condition without encapsulation ((i) device fabrication except the ambient exposure, RH35%; (ii) ambient exposure, RH45%; (iii) long-term storage, RH35%). Reproduced with permission from Ref. [34]. Copyright 2017, Nature Publishing Group.

In addition, carbon electrodes play a critical role in improving MPSCs' performance. Ku et al. firstly developed TiO_2/ZrO_2/carbon configuration PSCs device, in which spheroidal graphite was replaced by the flaky graphite based device, obtaining an improved PCE from 4.08% to 6.64% [26]. Afterwards, Xu and co-workers applied highly ordered mesoporous carbon with well-connected frameworks in MPSCs device. The mesoporous and interconnected structures favored the penetration of perovskite precursor solution and provided effective contact between $MAPbI_3$ and carbon thus improved the PCE of the device from 5.17% to 7.02% [38].

The impact of the graphite size in carbon electrode on the photovoltaic performance was studied by Zhang et al. [39]. It is found that the 8 mm graphite-based electrode possessed a larger average pore size, a smaller square resistance and hence a higher PCE exceeding 11%. Duan et al. employed mechanical exfoliated ultrathin graphite (UG) in MPSC devices. The UG effectively increased the specific surface area of the carbon layer without sacrificing the conductivity, as shown in Fig. 7.4a–c. The large specific surface area facilitated the hole collection from the perovskite to carbon layer, correspondingly decreased the charge transfer resistance at the perovskite/carbon interface. In addition, the penetration of $(5\text{-}AVA)_x(MA)_{1-x}PbI_3$ perovskite precursor solution in the carbon layer was accelerated due to the highly interconnected channels and perovskite grew well in most of the mesoporous. Accordingly, the PCE of the MPSC device was significantly improved from 12.63% to 14.07% by incorporating UG instead of bulk graphite [40].

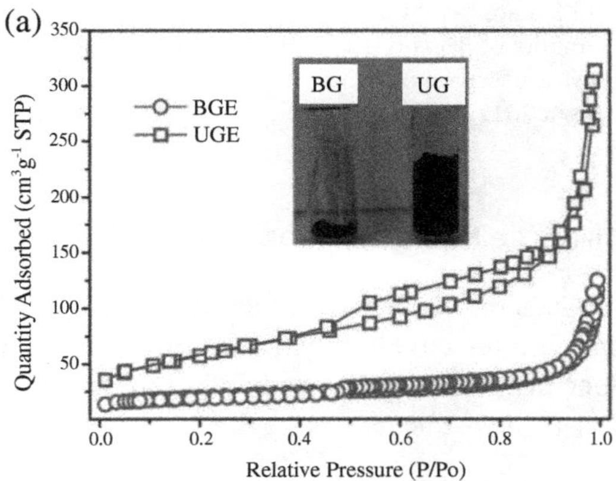

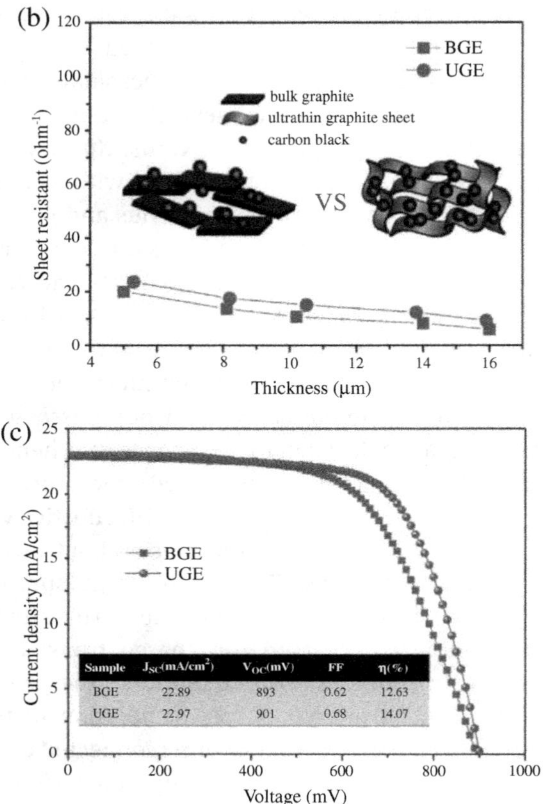

Figure 7.4 (a) Nitrogen gas adsorption and desorption isotherms of BGE and UGE. Digital photographs of 0.05 g BG and 0.05 g UG powders shown in the inset of the figure. (b) Square resistance of BGE and UGE films at different thickness and schematic model of BG and UG conductive network (inset). (c) *J–V* curves of the MPSC device based on BGE and UGE. Reproduced with permission from Ref. [40]. Copyright 2017, Elsevier.

7.3.2 HTM-Free Planar Carbon-Based PSCs

Panar carbon-based PSCs were first developed by Yang's group by directly clamping a selective hole extraction electrode made of candle soot [41]. Specifically, in the first attempt, candle soot was collected on FTO glass to give the necessary hole extraction functionality (Fig. 7.5).

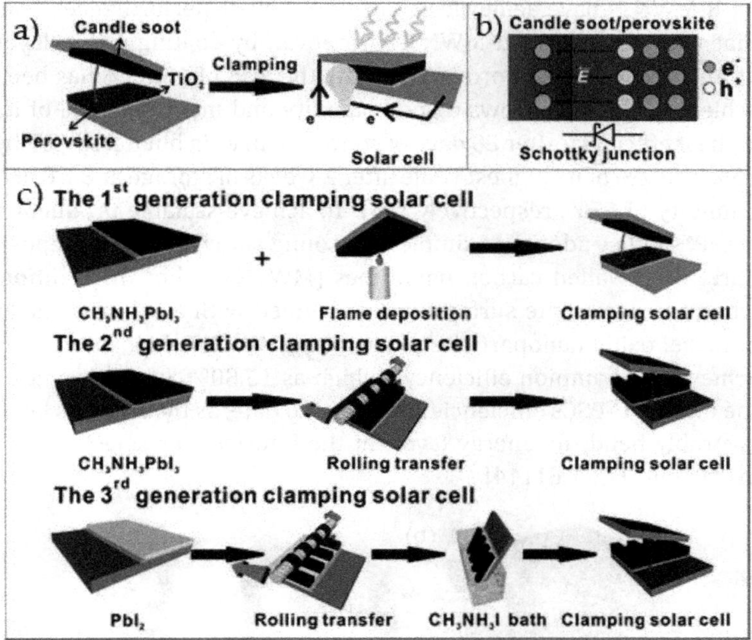

Figure 7.5 Development of a candle soot/perovskite clamping solar cell. Reproduced with permission from Ref. [41]. Copyright 2014, The Royal Society of Chemistry.

However, the resulted device performance was not good due to the low conductivity of the candle soot electrode. After the candle soot was replaced with a carbon black electrode (conductivity ≈ 5 S cm^{-1}) and greatly improved the device performance, the strategy of pre-wetting and reaction embedding the carbon particles was established to improve soot/perovskite interface contact. The champion device showed a PCE of 11.02% with a V_{OC} of 0.90 V, J_{SC} of 17.00 mA cm^{-2} and FF of 0.72.

Through using multi-walled carbon nanotube (MWCNT) as the counter electrode, they have achieved a hysteresis-free PCE of 12.67% with a very impressive FF of 0.80 due to the low-dimensional nanostructured and highly conductive characteristics [42]. After that, boron (B) doping of multiwalled carbon nanotubes (B-MWCNT) electrodes are designed, as it has superior in enabling enhanced hole extraction and transport by increasing work function, carrier concentration, and conductivity of MWNTs. The C-PSCs based on B-MWNTs have achieved remarkably higher performances than that with the undoped MWNTs. Moreover, by coating a thin layer of Al_2O_3 on the mesoporous TiO_2 film, the PCE of 15.23% has been achieved. The PCE shows a good stability and maintain 98% of its highest efficiency after 80 days of storage in dry air, but decreased by 15% and 7% of its highest value after 2 weeks of storage at 80°C and humidity of 65%, respectively [43]. To achieve scalable production of C-PSCs, they adopted a simple ultrasound spray method to deposit pure multi-walled carbon nanotubes (MWCNT) films. In addition, when the perovskite surface was embedded with a sub-monolayer of nickel oxide nanoparticles prior to the MWCNT deposition, they achieved a champion efficiency as high as 15.80%, which is among the highest C-PSCs efficiencies reported to date, as the NiO NPs layer favorably bend the energy levels at the interface for selective hole extraction (Fig. 7.6) [44].

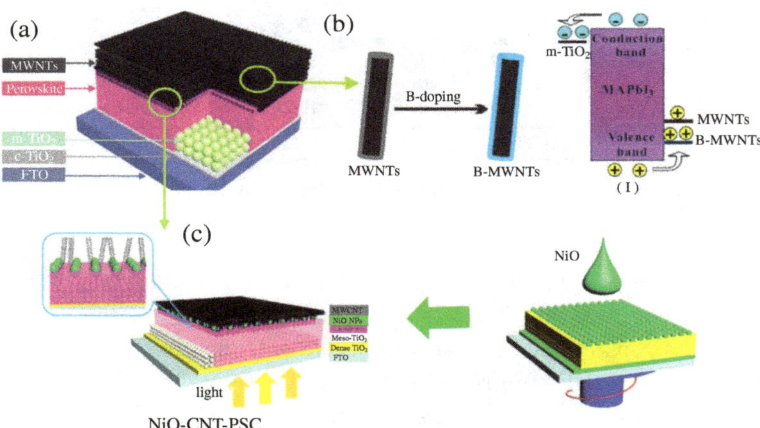

Figure 7.6 (a) Schematic illustration of cell configuration; (b) B-doping of MWNTs and band alignments of the solar cell; and (c) NiO nanoparticles-embedded solar cell. (a and b) reproduced with permission from Ref. [43]. Copyright 2017, American Chemical Society. (c) reproduced with permission from Ref. [44]. Copyright 2017, Elsevier.

In 2016, Liang et al. firstly reported carbon-based HTM-free all-Inorganic PSCs without any labile or expensive organic components. The used $CsPbBr_3$ to take the place of organic–inorganic hybrid perovskite and obtained a PCE of 6.7% with an active area of 0.12 cm^2 and 5.0%, with an active area of 1 cm^2. The relatively poor performance is due to the low photocurrent density, as the bandgap of $CsPbBr_3$ is as wide as 2.3 eV. However, due to the eliminating of organics, the PSCs present super stability. Specifically, the PSCs exhibited no degradation after storage in humid air (90–95% RH, 25°C) for over 3 months without any sealing. Additionally, the PCE of $CsPbBr_3$/carbon-based all-inorganic PSCs exhibited remarkable stability during a long testing period of 840 h in high temperature (100°C) and low temperature (–22°C). To simulate the real work condition of solar cells, the stability of $CsPbBr_3$/carbon-based all-inorganic PSCs during temperature cycles between –22 and 100°C were tested and the cells exhibited no degradation for 80 h without encapsulation [45]. After that, they adopted energy-band engineering strategy to tune the bandgap of $CsMX_3$ perovskite and improve the efficiency of the all-inorganic PSCs. By introduction and molar ratio adjusting of Sn and I elements, the $CsPb_{0.9}Sn_{0.1}IBr_2$ perovskite displays a suitable bandgap of 1.79 eV and an appropriate level of valence band maximum (Fig. 7.7). Appling this perovskite

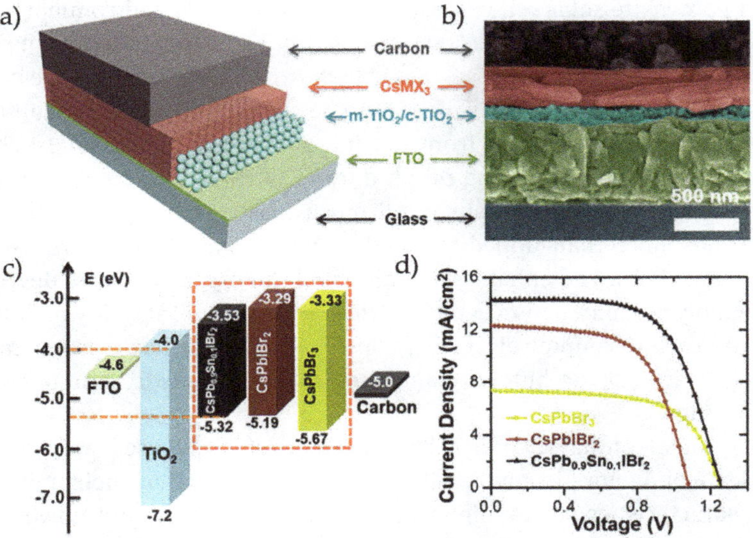

Figure 7.7 (a) Schematic view and (b) cross-sectional SEM image of all-inorganic PSCs with the configuration of FTO/c-TiO_2/m-TiO_2/$CsMX_3$/carbon. (c) Energy level diagrams of all-inorganic PSCs. The band levels of different $CsMX_3$ perovskites are shown in the red dashed-line box. (d) *J–V* plots of all-inorganic PSCs based on $CsPbBr_3$, $CsPbIBr_2$, and $CsPb_{0.9}Sn_{0.1}IBr_2$, respectively. Reproduced with permission from Ref. [46]. Copyright 2017, American Chemical Society.

into carbon-based HTM-free PSCs, the solar cell exhibits a high open-circuit voltage of 1.26 V and a remarkable PCE up to 11.33%, which is record-breaking among the existing $CsMX_3$-based PSCs. However, this new mixed-Pb/Sn mixed-halide perovskite has worse stability than $CsPbBr_3$, although superior than $CsPbI_3$. With encapsulation, the $CsPb_{0.9}Sn_{0.1}IBr_2$ carbon-based HTM-free PSCs exhibited little degradation after being kept for over 3 months at RT. They can also operate normally after being continuously heated at 100°C for >2 weeks but with some degradation. Without encapsulation, the PCE retains only 85% of its initial value after exposure to ambient air with 50–60% relative humidity at RT for 50 h [46].

7.3.3 Carbon-Based PSCs with HTMs

Due to the absence of HTM layer, perovskite and carbon layer will contact directly in HTM-free CPSCs, which could cause inefficient hole extraction, serious charge recombination and then drop off the performance of devices. To address the issue, Zhang et al. developed a perovskite solar cell with CuPc nano rod as HTM and commercial carbon as Counter electrode, as shown in Fig. 7.8a [47]. It is found that CuPc/Carbon based device showed an improved device performance with PCE of 16.1%. To be specific, during the stability tests, the PCEs change from the initial 16.1% to the final 14.7% with a 8.5% drop during 600 h durability tests. The CuPc/Carbon based PSC device demonstrated much better stability in ambient atmosphere than doped-spiro-OMeTAD/Au based device (shown in Fig. 7.8b). The enhanced durability of CuPc/carbon based device mainly attributed to the good thermal and chemical stability of CuPc and carbon counter electrode, which can also work as a protecting layer against the humidity and protect the $MAPbI_3$ from humidity damage.

Metal phthalocyanines have been extensively used as p-type semiconductors in photovoltaic field, due to some of their unique characteristics, such as high hole mobility, high stability, etc. However, commercially available copper phthalocyanine (CuPc) require a dedicated vacuum thermal evaporation step. Therefore, designing solution processable CuPc derivative to probe the fundamental principles that underlie their structure–perovskite performance is of utmost importance. In 2017, CuPc was molecularly engineered by the introduction of triisopropylsilylethynyl (TIPS) substituents in the periphery of the Pc ring to improve the solubility and hydrophobicity in Sun's group (shown in Fig. 7.9a) [48]. The substituent has been successfully adopted into pentacene derivatives to improve the solubility and molecule arrangements in the solid state. CuPc-TIPS was synthesized and explored as a HTM for PSCs, in combination with a mixed-ion perovskite absorber and a low-cost vacuum-free carbon counter electrode. The best PSC device based on pristine CuPc-TIPS showed a decent PCE of 14.0%. More strikingly, the devices studied also exhibited a good long-term stability under ambient conditions, due largely to the absence of hydroscopic lithium additives and the introduction of the hydrophobic HTM interlayer that prevented moisture from penetrating into the perovskite film, as shown in Fig. 7.9b.

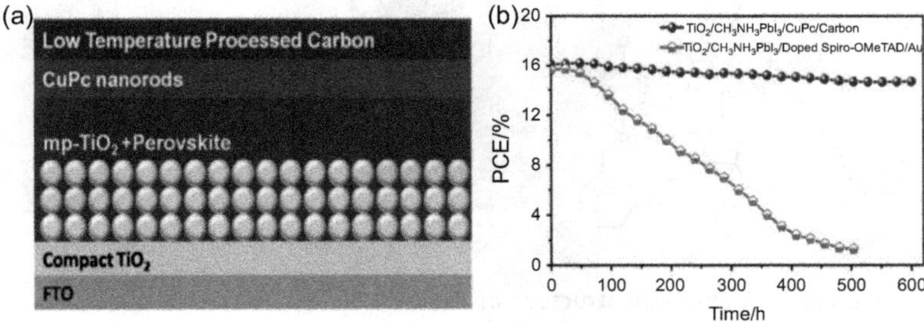

Figure 7.8 (a) Device architecture and energy-level diagram: schematics of the whole device: FTO glass/compact TiO_2 layer/mp-TiO_2/$MAPbI_3$ capping layer/CuPc/Carbon. (b) Stability tests of perovskite solar cells with CuPc/Carbon and doped–spiro-OMeTAD/Au as HTMs and counter electrodes respectively. Reproduced with permission from Ref. [47]. Copyright 2016, Elsevier.

Besides CuPc-TIPS, an organic–inorganic integrated hole transport layer (HTL) composed of the solution-processable nickel phthalocyanine (NiPc) abbreviated NiPc-$(OBu)_8$ and vanadium(V) oxide (V_2O_5) is incorporated into structured mesoporous perovskite solar cells, as shown in Fig. 7.9c–d [49]. Through careful interface modification, the devices containing NiPc-$(OBu)_8$ and V_2O_5 integrated hole transport layers show very competitive average PCEs of up to 17.6%. Moreover, the integrated HTL based devices exhibit better stability than doped Spiro-OMeTAD based ones. The present results open a new route to cost-effective and highly efficient

PSCs. These results highlight the potential application of organic–inorganic integrated HTLs in PSCs.

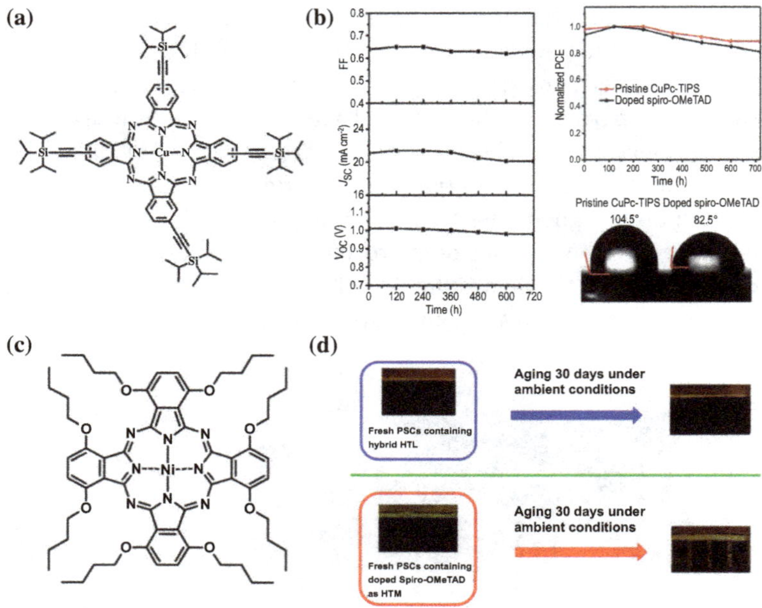

Figure 7.9 (a) Chemical structure of CuPc-TIPS. (b) Changes of photovoltaic parameters of V_{OC}, J_{SC}, and FF for PSCs based on pristine CuPc-TIPS, PCE variations of PSCs based on pristine CuPc-TIPS and doped spiro-OMeTAD, water contact angles of pristine CuPc-TIPS and spiro-OMeTAD doped with Li-TFSI and TBP thin films. (c) Chemical structure of NiPc-$(OBu)_8$. (d) Photographs of PSCs containing different HTLs before and after aging tests. (a and b) reproduced with permission from Ref. [48]. Copyright 2017, Royal Society of Chemistry. (c and d) reproduced with permission from Ref. [49]. Copyright 2017, John Wiley and Sons.

7.4 Scaling-Up of Carbon-Based Perovskite Solar Cells

For perovskite solar cells (PSCs) based on conventional structures, such as FTO/TiO_2/Perovskite/Spiro-OMeTAD/Au and ITO/PEDOT:PSS/Perovskite/PCBM/Al, it is challenging to enlarge the area of devices, due to the mostly used deposition process of spin-coating, and thermal evaporation [50–53]. Although various methods have been reported to fabricate large-area PSCs, the size of the device is still far from that for applications [54–57]. For carbon-based PSCs, especially for the triple mesoscopic layers structure carbon-based PSCs, it is much easier to enlarge the device area, since the fabrication process is mainly based on sprinting technology, and no vacuum condition is required. Besides, the replacement of noble metals as back contacts by low-cost carbon materials of graphite and carbon black further reduces the material cost of carbon-based PSCs [58, 59].

7.4.1 Architectures of Large-Area Carbon-Based PSC Modules

Large-area carbon-based PSC modules were firstly reported by Han group [28]. A series-connected architecture is developed for large-area carbon-based PSC modules to avoid ohmic losses. Figure 7.10a shows the digital image of a large-area carbon-based PSC module with the area of 100 cm^2, and Fig. 7.10b illustrates the architecture of the module. Ten sub-cells are constructed the FTO glass substrate in series interconnection with monolithic architecture. The overall active area is 49 cm^2, and the aperture area is 51 cm^2. Specifically, the length of the active area of a single cell is 94 mm, and the width of each sub-cell is 5.3 mm. For the fabrication of such modules, the FTO layer on the glass substrates is firstly etched by laser processing. Then compact-TiO_2 layer is sprayed on the patterned FTO glass substrates. The mesoporous TiO_2, ZrO_2 and carbon layer are screen printed on the compact-TiO_2 covered FTO substrates. The thickness of each mesoporous layer is controlled by modifying the mesh of the screen mask and the composition of the pastes. After sintering at 400–500°C for 30–40 mins, a perovskite precursor solution is deposited in the three mesoporous layers by drop-casting method. After drying at 50–70°C for 2–3 hours, the fabrication of a carbon-based PSC module is finished. Figures 7.10c and d show the energy level diagram and charge separation mechanism of the module. Upon illumination, the photogenerated electrons and holes are extracted from the perovskite layer to TiO_2 and carbon, and then to the front and back contact. The separated charge carriers can transport through the serially connected sub-cells and extract to the external circuit.

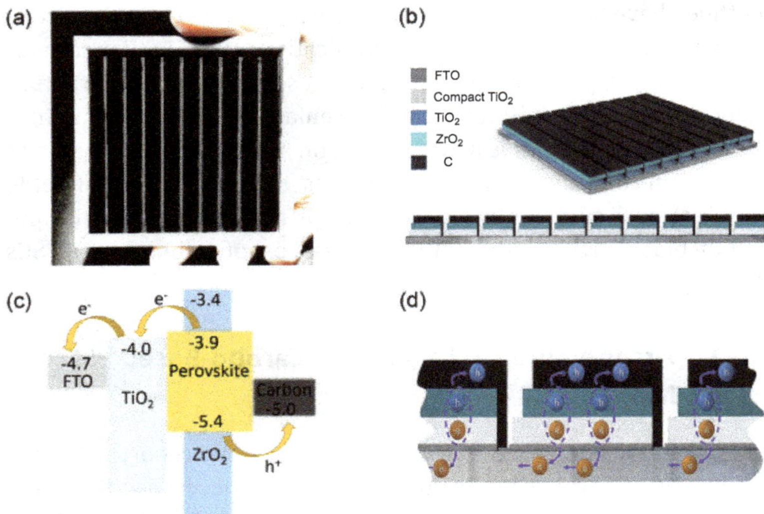

Figure 7.10 (a) Digital image of a typical large-area carbon-based PSC module with the area of 100 cm^2. (b) The scheme of the module which consists of 10 sub-cells in series interconnection. (c) Energy level diagrams of the carbon-based PSC module. (d) Schematic illustration of the electron and hole separation in a serially connected carbon-based PSC module. Reproduced with permission from Ref. [58]. Copyright 2017, John Wiley and Sons.

7.4.2 Printing Techniques

Screen printing is a widely used film deposition technique, for which a mesh (screen) is used to transfer ink/paste onto a flat substrate. The printed patterns are determined by the open mesh apertures of the screen. The non-printed areas are made impermeable to the ink/paste by a blocking stencil. The ink/paste is placed on the non-printed areas, and a printing squeegee is moved across the surface of the screen to fill the open mesh apertures with ink/paste. A reverse stroke of the squeegee then presses the screen to contact the substrate. This causes the ink to wet the substrate and be pulled out of the mesh apertures as the screen springs back after the squeegee has passed, as shown in Fig. 7.11.

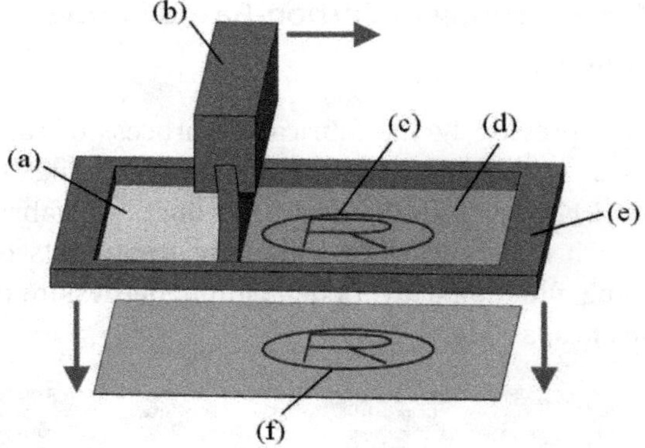

Figure 7.11 The scheme of screen-printing technique. The components are labeled as (a) ink/paste, (b) squeegee, (c) open mesh aperture of the screen, (d) impermeable area of the screen, (e) frame of the screen, (f) printed patterns on the substrate.

Typically, the screen is made of a piece of mesh stretched over a metallic frame. The mesh could be made of a synthetic polymer, such as nylon, and must be fixed on a frame under specific tension. The frame which holds the mesh could be made of diverse materials, such as aluminum.

The screen is usually placed on top of the substrate. Ink/paste is placed on the screen, and a scraper is used to push the ink/paste through the holes in the mesh. The operator begins with the scraper at the rear of the screen and behind a reservoir of ink/paste. The operator lifts the screen to prevent contact with the substrate and then using a slight amount of downward force pulls the scraper to the front of the screen. This effectively fills the mesh openings with ink/paste and moves the ink reservoir to the front of the screen. The operator then uses a squeegee (rubber blade) to move the mesh down to the substrate and pushes the squeegee to the rear of the screen. The ink/paste that is in the mesh opening is pumped or squeezed by capillary action to the substrate in a controlled and prescribed amount, i.e. the wet ink deposit is proportional to the thickness of the mesh and or stencil. As the squeegee moves toward the rear of the screen the tension of the mesh pulls the mesh up away from the substrate (called snap-off) leaving the ink upon the substrate surface.

7.4.3 Manufacturing of Carbon-Based Perovskite Solar Modules

As mentioned previously, the fabrication process of carbon-based PSC modules is mainly based on screen-printing techniques [58]. Figure 7.12 shows the typical procedures for fabricating the modules, which mainly includes substrate treatments of cleaning and patterning, functional layers deposition, perovskite deposition, and module encapsulation.

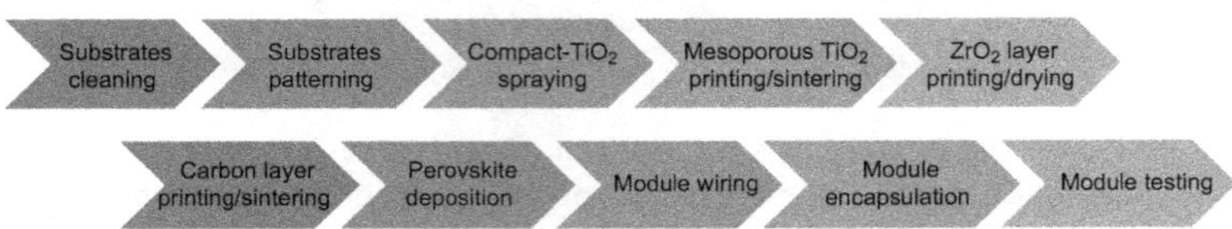

Figure 7.12 The typical fabrication procedures of carbon-based perovskite solar modules.

Firstly, the FTO layer on the glass substrate is etched by a laser beam with a wavelength of 1064 nm. Then the patterned substrates are ultrasonically cleaned by detergent, deionized water and ethanol. The substrates are then heated up to 450°C, and a compact layer, such as TiO_2 and Al_2O_3, is sprayed. When the compact layer covered substrates cool down to room temperature, a mesoporous TiO_2 layer is screen printed and sintered at 500°C for 40 min. After cooling down to room temperature, a mesoporous ZrO_2 layer is screen printed on the TiO_2 layer and dried at 70°C for 40 min. Then a carbon layer is screen printed on the ZrO_2 layer and sintered at 400°C for 30 min. Here the construction of the mesoporous scaffold of TiO_2/ZrO_2/Carbon triple-layer is finished. A perovskite precursor solution is prepared and kept stirring at 50°C. For the small-area lab cells, the perovskite is deposited in the mesoporous scaffold with a simple drop-casting method. For the large-area modules, the perovskite precursor solution is deposited in the mesoporous scaffold by an automatic syringe, inkjet printing, or slot die. After drying at 50–70°C for 2–3 hours, wires are connected to the front and back contacts. The fabrication of the module is almost finished. For the encapsulation of PSCs, there have been many options, such as using hot melt films and so on [60, 61]. Although it is yet unclear whether the different encapsulation methods perform equally well or if certain methods will prove beneficial over others, the encapsulation absolutely can prolong the lifetime of the devices and modules. Recently, a stability test with a period of over 10000 hours has been performed, which provides a promising prospect for further wide applications [31].

The advantages of carbon-based perovskite solar modules mainly lie in the simple fabrication process based on screen-printing techniques and low-cost materials. Particularly, it is straightforward to enlarge the module area by using a large screen. Recently, WonderSolar Co. Ltd. [62] in China developed a printable carbon-based perovskite solar module with the area of 3600 cm^2, and proposed a continuous production line for such modules, as shown in Fig. 7.13. With a module dimension of 60 cm × 120 cm, the production line can achieve a capacity of over 50 MW with an investment of only 15~30 million RMB.

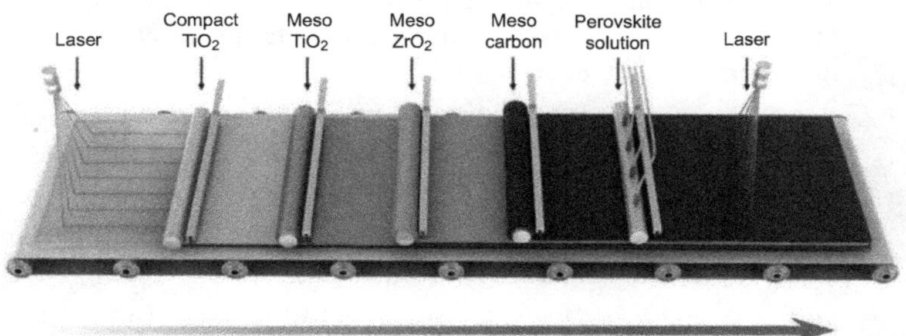

Figure 7.13 Schematic illustration of the proposed production line of carbon-based PSC modules. Reproduced with permission from Ref. [48]. Copyright 2017, John Wiley and Sons.

7.5 Conclusions

Perovskite solar cells have undergone a number of changes since their initial implementation eight year ago. Among all types of PSCs, carbon-based PSCs have been paid much attention to from both scientists and industries. The advantages such as low-cost, easy-fabrication and high stability make carbon-based PSC a fantastic candidate for large-scale applications and commercialization of PSCs. It is important to recognize that the general photovoltaics context has also radically altered and progressed since the first PSCs were developed. The cost of silicon photovoltaics has plummeted, while other types of photovoltaics are also searching for their purpose and opportunities such as indoor technology for powering the internet of things, building-integrated photovoltaics and so on. The priority and great emphasis put on carbon-based PSCs remains improving the PCE under the condition of low-cost. By exploiting innovative strategies from different aspects, such as materials, interfaces, device structure, stability measurement consistency and so on, we believe that the carbon-based PSCs could still be further promoted and pave the way for large-scale commercial deployment of this very promising photovoltaic technology.

References

1. Kojima, A., Teshima, K., Shirai, Y., and Miyasaka, T. (2009). Organometal halide perovskites as visible-light sensitizers for photovoltaic cells. *J. Am. Chem. Soc.,* **131**, 6050–6051.
2. Yang, W. S., et al. (2017). Iodide management in formamidinium-lead-halide–based perovskite layers for efficient solar cells. *Science,* **356**, 1376.
3. Green, M. A., et al. (2017). Solar cell efficiency tables (version 50). *Prog. Photovoltaics Res. Appl.,* **25**, 668–676.
4. Yin, W.-J., Yang, J.-H., Kang, J., Yan, Y., and Wei, S.-H. (2015). Halide perovskite materials for solar cells: a theoretical review. *J. Mater. Chem. A,* **3**, 8926–8942.
5. Ibn-Mohammed, T., et al. (2017). Perovskite solar cells: an integrated hybrid lifecycle assessment and review in comparison with other photovoltaic technologies. *Renewable Sustainable Energy Rev.,* **80**, 1321–1344.
6. Kim, H.-S., et al. (2012). Lead iodide perovskite sensitized all-solid-state submicron thin film mesoscopic solar cell with efficiency exceeding 9%. *Sci. Rep.,* **2**, 591.
7. Lee, M. M., Teuscher, J., Miyasaka, T., Murakami, T. N., and Snaith, H. J. (2012). Efficient hybrid solar cells based on meso-superstructured organometal halide perovskites. *Science,* **338**, 643–647.
8. Shin, S. S., et al. (2017). Colloidally prepared La-doped $BaSnO_3$ electrodes for efficient, photostable perovskite solar cells. *Science,* **356**, 167.

9. Park, N.-G., Grätzel, M., Miyasaka, T., Zhu, K., and Emery, K. (2016). Towards stable and commercially available perovskite solar cells. *Nat. Energy*, **1**, 16152.

10. Saliba, M., et al. (2016). Incorporation of rubidium cations into perovskite solar cells improves photovoltaic performance. *Science*, **354**, 206–209.

11. Yang, W. S., et al. (2015). High-performance photovoltaic perovskite layers fabricated through intramolecular exchange. *Science*, **348**, 1234–1237.

12. Dong, Q., et al. (2015). Electron-hole diffusion lengths >175 μm in solution-grown $CH_3NH_3PbI_3$ single crystals. *Science*, **347**, 967–970.

13. Cai, M., et al. (2017). Cost-performance analysis of perovskite solar modules. *Adv. Sci.*, **4**.

14. Berhe, T. A., et al. (2016). Organometal halide perovskite solar cells: degradation and stability. *Energy Environ. Sci.*, **9**, 323–356.

15. Rong, Y., Liu, L., Mei, A., Li, X., and Han, H. (2015). Beyond efficiency: the challenge of stability in mesoscopic perovskite solar cells. *Adv. Energy Mater.*, **5**.

16. Tzounis, L., et al. (2017). Perovskite solar cells from small scale spin coating process towards roll-to-roll printing: optical and morphological studies. *Mater. Today.*, **4**, 5082–5089.

17. Williams, S. T., Rajagopal, A., Chueh, C.-C., and Jen, A. K.-Y. (2016). Current challenges and prospective research for upscaling hybrid perovskite photovoltaics. *J. Phys. Chem. Lett.*, **7**, 811–819.

18. Hailegnaw, B., Kirmayer, S., Edri, E., Hodes, G., and Cahen, D. (2015). Rain on methylammonium lead iodide based perovskites: possible environmental effects of perovskite solar cells. *J. Phys. Chem. Lett.*, **6**, 1543–1547.

19. Abate, A. (2017). Perovskite solar cells go lead free. *Joule*, **1**, P659–P664.

20. Babayigit, A., Ethirajan, A., Muller, M., and Conings, B. (2016). Toxicity of organometal halide perovskite solar cells. *Nat. Mater.*, **15**, 247–251.

21. Herz, L. M. (2016). Charge-carrier dynamics in organic-inorganic metal halide perovskites. *Annu. Rev. Phys. Chem.*, **67**, 65.

22. Hu, Q., et al. (2017). In situ dynamic observations of perovskite crystallisation and microstructure evolution intermediated from [PbI6](4-) cage nanoparticles. *Nat. Commun.*, **8**, 15688.

23. Saliba, M., Correa-Baena, J.-P., Graetzel, M., Hagfeldt, A., and Abate, A. (2017). Perovskite solar cells from the atomic to the film level. *Angew. Chem. Int. Ed.*, **57**.

24. Hu, R., Chu, L., Zhang, J., Li, X. A., and Huang, W. (2017). Carbon materials for enhancing charge transport in the advancements of perovskite solar cells. *J. Power Sources*, **361**, 259–275.

25. McCreery, R. L. (2008). Advanced carbon electrode materials for molecular electrochemistry. *Chem. Rev.*, **108**, 2646–2687.

26. Ku, Z., Rong, Y., Xu, M., Liu, T., and Han, H. (2013). Full printable processed mesoscopic $CH_3NH_3PbI_3/TiO_2$ heterojunction solar cells with carbon counter electrode. *Sci. Rep.*, **3**, 3132.

27. Mei, A., et al. (2014). A hole-conductor–free, fully printable mesoscopic perovskite solar cell with high stability. *Science*, **345**, 295–298.

28. Hu, Y., et al. (2017). Stable large-area ($10 \times 10\,cm^2$) printable mesoscopic perovskite module exceeding 10% efficiency. *Solar RRL*, **1**.

29. Jiang, X., et al. (2017). High-performance regular perovskite solar cells employing low-cost poly (ethylenedioxythiophene) as a hole-transporting material. *Sci. Rep.*, **7**.

30. Baranwal, A. K., et al. (2016). 100°C thermal stability of printable perovskite solar cells using porous carbon counter electrodes. *ChemSusChem*, **9**, 2604–2608.

31. Grancini, G., et al. (2017). One-Year stable perovskite solar cells by 2D/3D interface engineering. *Nat. Commun.*, **8**.

32. Hong, L., et al. (2017). Improvement and regeneration of perovskite solar cells via methylamine gas post-treatment. *Adv. Funct. Mater.*, **27**.

33. Hou, X., et al. (2016). Effect of guanidinium on mesoscopic perovskite solar cells. *J. Mater. Chem. A*, **5**.

34. Rong, Y., et al. (2017). Synergy of ammonium chloride and moisture on perovskite crystallization for efficient printable mesoscopic solar cells. *Nat. Commun.*, **8**, 14555, doi:10.1038/ncomms14555.

35. Chen, J., et al. (2016). Solvent effect on the hole-conductor-free fully printable perovskite solar cells. *Nano Energy*, **27**, 130–137, doi:http://dx.doi.org/10.1016/j.nanoen.2016.06.047.

36. Chen, J., et al. (2016). Hole-conductor-free fully printable mesoscopic solar cell with mixed-anion perovskite $CH_3NH_3PbI_{(3-x)}(BF4)_x$. *Adv. Energy Mater.*, **6**.

37. Sheng, Y., et al. (2016). Enhanced electronic properties in $CH_3NH_3PbI_3$ via LiCl mixing for hole-conductor-free printable perovskite solar cells. *J. Mater. Chem. A*, **4**, 16731–16736.

38. Xu, M., et al. (2014). Highly ordered mesoporous carbon for mesoscopic $CH_3NH_3PbI_3/TiO_2$ heterojunction solar cell. *J. Mater. Chem. A,* **2**, 8607–8611.

39. Zhang, L., et al. (2015). The effect of carbon counter electrodes on fully printable mesoscopic perovskite solar cells. *J. Mater. Chem. A,* **3**, 9165–9170.

40. Duan, M., et al. (2017). Efficient hole-conductor-free, fully printable mesoscopic perovskite solar cells with carbon electrode based on ultrathin graphite. *Carbon,* **120**, 71–76.

41. Wei, Z., et al. (2014). Cost-efficient clamping solar cells using candle soot for hole extraction from ambipolar perovskites. *Energy Environ. Sci.,* **7**, 3326–3333.

42. Wei, Z., Chen, H., Yan, K., Zheng, X., and Yang, S. (2015). Hysteresis-free multi-walled carbon nanotube-based perovskite solar cells with a high fill factor. *J. Mater. Chem. A,* **3**, 24226–24231.

43. Zheng, X., et al. (2017). Boron doping of multiwalled carbon nanotubes significantly enhances hole extraction in carbon-based perovskite solar cells. *Nano Lett.,* **17**, 2496–2505.

44. Yang, Y., et al. (2017). Ultrasound-spray deposition of multi-walled carbon nanotubes on NiO nanoparticles-embedded perovskite layers for high-performance carbon-based perovskite solar cells. *Nano Energy,* **42**, 322–333.

45. Liang, J., et al. (2016). All-inorganic perovskite solar cells. *J. Am. Chem. Soc.,* **138**, 15829–15832.

46. Liang, J., et al. (2017). $CsPb_{0.9}Sn_{0.1}IBr_2$ based all-inorganic perovskite solar cells with exceptional efficiency and stability. *J. Am. Chem. Soc.,* **139**, 14009–14012.

47. Zhang, F., Yang, X., Cheng, M., Wang, W., and Sun, L. (2016). Boosting the efficiency and the stability of low cost perovskite solar cells by using CuPc nanorods as hole transport material and carbon as counter electrode. *Nano Energy,* **20**, 108–116.

48. Jiang, X., et al. (2017). A solution-processable copper (II) phthalocyanine derivative as a dopant-free hole-transporting material for efficient and stable carbon counter electrode-based perovskite solar cells. *J. Mater. Chem. A,* **5**, 17862–17866.

49. Cheng, M., et al. (2017). Efficient perovskite solar cells based on a solution processable nickel (II) phthalocyanine and vanadium oxide integrated hole transport layer. *Adv. Energy Mater.,* **7**.

50. Kim, H.-S., et al. (2012). Lead iodide perovskite sensitized all-solid-state submicron thin film mesoscopic solar cell with efficiency exceeding 9%. *Sci. Rep.,* **2**, 591.

51. You, J., et al. (2014). Low-temperature solution-processed perovskite solar cells with high efficiency and flexibility. *ACS Nano,* **8**, 1674–1680.

52. Bi, D., et al. (2016). Polymer-templated nucleation and crystal growth of perovskite films for solar cells with efficiency greater than 21%. *Nat. Energy,* **1**, 16142, doi:10.1038/nenergy.2016.142 https://www.nature.com/articles/nenergy2016142#supplementary-information.

53. Xiao, Z. G., et al. (2014). Efficient, high yield perovskite photovoltaic devices grown by interdiffusion of solution-processed precursor stacking layers. *Energy Environ. Sci.,* **7**, 2619–2623, doi:10.1039/c4ee01138d.

54. Chen, W., et al. (2015). Efficient and stable large-area perovskite solar cells with inorganic charge extraction layers. *Science* **350**, 944–948, doi:10.1126/science.aad1015.

55. Li, X., et al. (2016). A vacuum flash–assisted solution process for high-efficiency large-area perovskite solar cells. *Science,* **353**, 58–62.

56. Ye, F., et al. (2016). Soft-cover deposition of scaling-up uniform perovskite thin films for high cost-performance solar cells. *Energy Environ. Sci.,* **9**, 2295–2301, doi:10.1039/C6EE01411A.

57. Bu, T., et al. (2017). A novel quadruple-cation absorber for universal hysteresis elimination for high efficiency and stable perovskite solar cells. *Energy Environ. Sci.*, doi:10.1039/C7EE02634J.

58. Hu, Y., et al. (2017). Stable large-area ($10 \times 10\,cm^2$) printable mesoscopic perovskite module exceeding 10% efficiency. *Solar RRL,* **1**, 1600019, doi:10.1002/solr.201600019.

59. Priyadarshi, A., et al. (2016). A large area (70 cm^2) monolithic perovskite solar module with a high efficiency and stability. *Energy Environ. Sci.,* **9**, 3687–3692, doi:10.1039/C6EE02693A.

60. Cheacharoen, R., et al. (2018). Design and understanding of encapsulated perovskite solar cells to withstand temperature cycling. *Energy Environ. Sci.*, doi:10.1039/C7EE02564E.

61. Han, Y., et al. (2015). Degradation observations of encapsulated planar $CH_3NH_3PbI_3$ perovskite solar cells at high temperatures and humidity. *J. Mater. Chem. A,* **3**, 8139–8147.

62. www.wondersolar.cn.

Chapter 8

Halide Perovskite Light-Emitting Diodes

Young-Hoon Kim, Soyeong Ahn, Joo Sung Kim, and Tae-Woo Lee
Department of Materials Science and Engineering, Seoul National University, 1 Gwanak-ro, Gwanak-gu, Seoul 08826, Republic of Korea
twlees@snu.ac.kr

8.1 Introduction

Organic–inorganic halide perovskites (OIPs) are promising light emitters due to their advantages such as high color purity with narrower light spectrum (full width at half maximum (FWHM) ≤ 20 nm) than that of organic emitters (FWHM ≥ 40 nm) and inorganic quantum dot (QD) emitters (FWHM ~ 30 nm), simple color tunability, and photoluminescence (PL) wavelength and color purity that are independent of grain size or particle size [1–4]. Furthermore, OIPs with ABX_3 or A_2BX_4 structure, where A is an organic ammonium (e.g., methylammonium (MA; $CH_3NH_3^+$), formamidinium (FA; $CH(NH_2)_2^+$)); B is a transition metal cation (e.g., Pb^{2+}, Sn^{2+}); X is a halogen anion (Cl, Br^-, I^-)) combine the

Multifunctional Organic–Inorganic Halide Perovskite: Applications in Solar Cells, Light-Emitting Diodes, and Resistive Memory
Edited by Nam-Gyu Park and Hiroshi Segawa

ISBN 978-981-4800-52-5 (Hardcover), 978-1-003-27593-0 (eBook)
www.jennystanford.com

advantageous properties of inorganic materials (e.g., high charge mobility) and organic materials (e.g., solution processibility, low cost, similar energy levels with organic materials) [1–7].

OIPs were developed as light emitters for light-emitting diodes (LEDs) earlier than as light absorbers for photovoltaics (PVs). In the early 1990s, OIPs with A_2BX_4 structure were used as an emission layer in LEDs. However, OIPs showed bright electroluminescence (EL) only at cryogenic temperature [5] or too low EL intensity to quantify the EL efficiency at room temperature (RT) [6, 7]. Thus, until 2013, researchers concentrated on organic emitters [8–13] and inorganic QD emitters [14–20] rather than OIP emitters to demonstrate high-efficiency LEDs.

In 2014, the bright EL of OIP polycrystalline (PC) bulk films at RT was reported [2] and promoted researchers to focus on increasing the EL efficiency of OIP PC films. To increase the luminescence efficiency (LE) of OIP LEDs, researchers have used the opto-electronic information of OIPs that was learned in studies of OIP PVs [21–24] and technical skills that they acquired from research on organic LED and inorganic QD LEDs [8–20]. Consequently, the LE of OIP LEDs have been increased significantly; examples include external quantum efficiency (EQE) ~ 8.53% and CE ~ 42.9 cd/A for green emitting OIP LEDs [4, 25], and EQE ~ 11.7% and radiance ~ 82 W sr^{-1} m^{-2} in near-infrared (NIR) emitting OIP LEDs [26].

All-inorganic halide perovskite (IHP) PC films in which A site in ABX_3 structure is an alkali metal cation such as Cs^+ have shown significant increases in LE (e.g., EQE ~ 4.26% and CE ~ 15.67 cd/A in green-emitting PeLEDs based on $CsPbBr_3$ PC [27]; estimated EQE ~ 10.435% using Lambertian assumption) and CE ~ 33.9 cd/A in green-emitting PeLEDs based on $Cs_{0.87}MA_{0.13}PbBr_3$ PC [28]). These increases of LE in PeLEDs (including OIP LEDs and IHP LEDs) are much faster than the progress of organic LEDs and inorganic QD LEDs and thus, show great possibility that OIPs and IHPs will be primary emitters in future displays and solid-lighting technology (Fig. 8.1).

In this chapter, we review the latest research into development of uniform and bright OIP and IHP PC bulk film emitters, and of high-efficiency PeLEDs based on them. In Section 8.2, we introduce device structure and working mechanism of PeLEDs, optoelectrical properties of perovskite PC films, and how perovskite PC films are fabricated and modified to improve their luminescent properties. In Section 8.3, we provide the perovskite crystal growth and nucleation mechanism in one-step solution process, and describe various strategies to fabricate uniform and bright perovskite PC films and to improve the LE of PeLEDs. In Section 8.4, we explain the perovskite crystallization mechanism in two-step solution process and introduce PeLEDs fabricated by using two-step solution process. In Section 8.5, we will give a perspective for PeLEDs and conclude. We provide better understanding of the properties of OIP and IHP PC film emitters, and progress in their development.

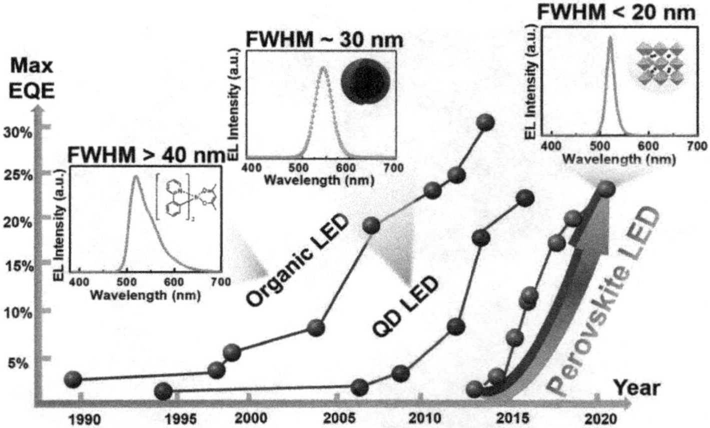

Figure 8.1 Development progresses of display technology. Modified from Ref. [1] with permission from National Academy of Sciences.

8.2 Fundamental Properties

8.2.1 Structure and Working Mechanism of PeLEDs

PeLEDs have similar device structure and working mechanism with OLEDs and inorganic QD LEDs; they are composed of anode, hole injection layer (HIL), perovskite emission layer, electron injection layer (EIL), and cathode (Fig. 8.2a) [2, 3]. When the external bias is applied to the device, holes and electrons are injected from anode to the HIL and from cathode to EIL, respectively. Highest occupied molecular orbital (HOMO) energy level of HIL locates between valence band maximum (VBM) energy level of perovskite and work function (WF) of anode and thus, facilitates the hole injection into perovskite emissive layer. Lowest unoccupied molecular orbital (LUMO) energy level of EIL located between the conduction band minimum (CBM) energy level of perovskite and WF of cathode and also promote the electron injection into perovskite layer. Injected electron and holes recombine in the perovskite emissive layer and make photons which have energy similar to the bandgap of perovskite emissive layer. Generated photons are emitted through the transparent electrode as a light.

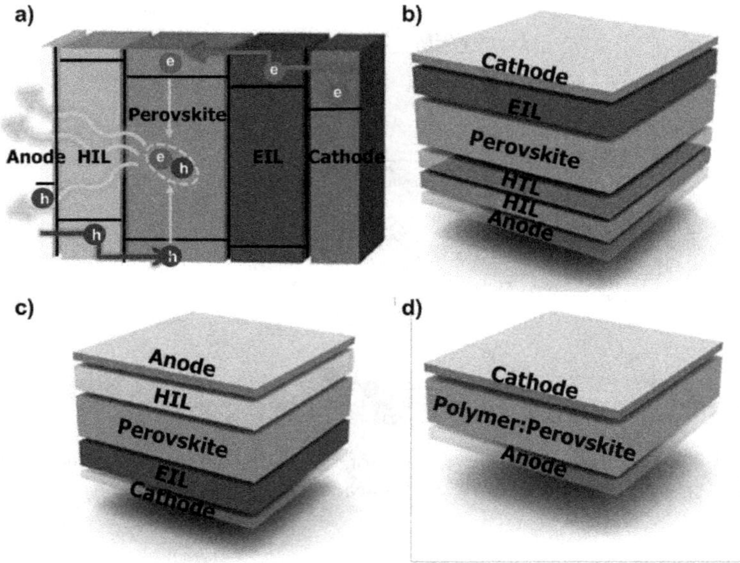

Figure 8.2 (a) Device structure and operation mechanism of conventional PeLEDs, and device structure of PeLEDs with (b) HTL interlayers, (c) inverted structure, and (d) simplified structure.

Hole transport layer (HTL) or electron transport layer (ETL) further facilitate charge injection into perovskite emission layer and prevent quenching of charge carriers at the HIL/perovskite or EIL/perovskite interfaces (Fig. 8.2b). PeLEDs with inverted structure in which electrons are injected from the bottom electrode and holes are injected from the top electrode of the devices have also been reported (Fig. 8.2c) [2]. Furthermore, PeLEDs with highly simplified structure (anode/emissive layer/cathode) can be demonstrated by fabricating ionic conductor polymer:perovskite composite films as an emissive layer (Fig. 8.2d) [29, 30].

8.2.2 Charge Carrier Dynamics of Perovskite Emitters

Behaviors such as recombination of excited charge carriers are important factors that determine the LE of perovskite PC films and PeLEDs. After perovskite PC films are excited by photo-illumination, electrons in the VBM are excited to the CBM; as a result, geminate electron-hole pairs are generated. They can (1) directly recombine to emit light (radiative recombination of excitons) or (2) separate into free carriers, which can then (2.1) radiatively recombine (radiative recombination of free charge carriers) or (2.2) become trapped in the defects (Shockley–Read–Hall (SRH) recombination), or (2.3) recombine by three-body Auger recombination (Fig. 8.3) [1, 31, 32].

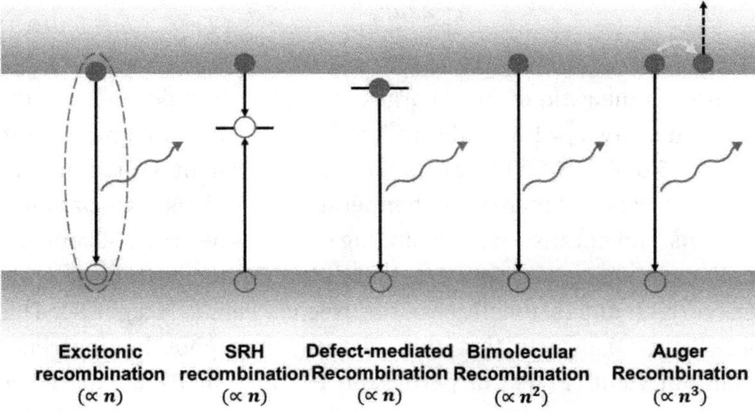

Figure 8.3 Recombination pathways of charge carriers.

The charge carrier dynamics in perovskite emitters can be described mainly by exciton binding energy. An exciton is an electron-hole pair that is bound by electrostatic coulombic force [33, 34]. An exciton has effective mass and binding energy. The exciton radius can be achieved as

$$a_0^* = \varepsilon_r \left(\frac{m_0}{\mu} \right) a_0 , \tag{8.1}$$

where ε_r is the relative dielectric constant, a_0 = 0.529 Å is the Bohr radius, m_0 is the mass of a free electron, $\frac{1}{\mu}=\frac{1}{m_e}+\frac{1}{m_h}$ is the reduced carrier mass where m_e is the effective mass of an electron and m_h is the effective mass of a hole. The exciton binding energy is

$$E_B = 13.6\frac{\mu}{m_0\varepsilon_r^2}, \tag{8.2}$$

where the constant comes from definition of the Rydberg (1 Ry = 13.6 eV). Excitonic behavior of perovskite PC films with large grain size or high dimensionality at RT are generally claimed to resemble Wannier-Mott excitons that have low exciton binding energy between 5–150 meV, rather than Frenkel excitons [35–41].

Considering free carriers and excitons in excited states, the number of carriers in each state can be described by the Saha equation [42, 43]:

$$\frac{x^2}{1-x}=\frac{1}{n}\left(\frac{2\pi\mu k_B T}{h^2}\right)^{3/2} e^{-\frac{E_B}{k_B T}}, \tag{8.3}$$

where x is the ratio of free charges to total exciton density, n is the charge density, $k_B = 1.38\times10^{23}\ m^2kgs^{-2}K^{-1}$ is the Boltzmann constant, and $h = 6.626 \times 10^{-34}\ m^2kg/s$ is the Planck constant. Equation (8.3) demonstrates that increase in temperature promotes dissociation of excitons, and because exciton binding energy is between 50 and 150 meV, most charge carriers in perovskite PC films with large grain size or high dimensionality exist as free carriers ($x \approx 1$) at RT. This observation suggests that the relatively low photoluminescence quantum yield (PLQY) of perovskite PC films at RT compared to that at low temperature can be attributed to exciton dissociation at elevated temperature [2]. Formation of optical phonons and polarons may be another reason for this low PLQY [36, 44]; this idea will not be discussed in this chapter.

To investigate precise radiative recombination kinetics on perovskite emitters, carrier dynamics can be fitted to a conventional recombination model as [45, 46]:

$$\frac{\partial n(t)}{\partial t}=-k_1 n-k_2 n^2-k_3 n^3, \tag{8.4}$$

where $n(t)$ is charge density dependent on time, k_1 is the monomolecular recombination coefficient of excitonic radiative recombination or defect-mediated recombination, k_2 corresponds to free-carrier radiative recombination, and k_3 corresponds to three-body Auger recombination.

The PLQY $\eta(n)$ of perovskite emitters can be described as [45, 46]:

$$\eta(n) = \frac{k_1' + k_1'' + k_2 n}{k_1' + k_{\text{trap}} + k_2 n + k_3 n^2}, \tag{8.5}$$

In (8.5), k_1 from Eq. (8.4) is partitioned into radiative excitonic term k'_1, trap-assisted radiative term k_1'' and defect-mediated nonradiative SRH term k_{trap}. k_1', k_1'' and bimolecular radiative recombination term $k_2 n$ can contribute to PLQY. k_1' and $k_2 n$ terms from Eq. (8.5) mainly contribute to PLQY of perovskite PC films, because (1) bimolecular recombination from free carriers is the dominant radiative recombination mechanism in perovskite PC films with large grain size or high dimensionality, (2) excitonic radiative recombination can contribute to PLQY in perovskite PC films with small grain size or low dimensionality and (3) trap-assisted recombination occurs mainly by nonradiative recombination [47, 48]. PLQY continuously increases with excitation density to ~ 50% in three dimensional (3D) perovskite PC bulk films and to > 60% in low-dimensional perovskite PC bulk films, because excitonic radiative recombination and bimolecular recombination become dominant over trap-assisted SRH recombination as trap sites are filled [39, 48]. Photoluminescence quantum yield (PLQY) is higher in lower-dimensional perovskite than in higher-dimensional carriers because low-dimensional perovskites more efficiently confine charge carriers; this effect indicates that decrease of dimensionality or size of perovskite grains can induce the higher probability of excitonic radiative recombination in confined system and thus, improve the PLQY of perovskite PC films and the LE of PeLEDs. However, at extremely high injected charge carrier density (n), PLQY decreases as n increases; the trend is due to increasing contribution of three-body Auger recombination, and this increase may occur because of exciton dissociation due to the interaction between excitons and charge-screening effects [48].

Especially in 3D perovskite PC films, k_2 ($\sim 10^{-10}$ cm^{-3} s^{-1}) and k_3 ($\sim 10^{27}$ cm^{6} s^{-1}) are intrinsic parameters [31, 49–52]. Thus, reducing trap-mediated recombination rate (k_{trap} in Eq. (8.5)) by decreasing trap density in perovskite PC films can be one strategy to achieve high PLQY in perovskite PC films [48, 53]. Therefore, to achieve high LE in PeLEDs, various methods to increase excitonic radiative recombination by decreasing grain size or dimensionality in perovskite crystal, and improve the morphology and reduce the trap density of perovskite PC films have been tried.

8.2.3 Crystallization Mechanism of Perovskite PC Bulk Films

Surface morphology of perovskite PC films and grain size of perovskite crystals are important factors which determine LE of PeLEDs. Perovskite PC film formation mechanism follows the crystallization kinetics. Thus, this section will present thermodynamic principles of perovskite crystallization and factors that control crystallization.

Perovskite crystallization kinetics can be described by the general thermodynamic principle [54, 55]:

$$\Delta G = V\Delta G_v + S\Delta G_s, \tag{8.6}$$

where ΔG is the change of Gibbs free energy, ΔG_V is the change of Gibbs free energy per unit volume, ΔG_S is the change of Gibbs free energy per unit area, V is the volume of crystal and S is its surface area. To fabricate perovskite PC films by crystallization, MAX precursors collide with PbX_2 precursors according to the probabilistic model. Perovskite nuclei are collided MAX and PbX_2 that have energy higher than activation energy; they begin to grow following the thermodynamic model. ΔG_V can be described as (8.7) [54–56]:

$$\Delta G_{\text{V}} = -\frac{kT}{V_{\text{m}}}\ln\frac{C}{C_0(T)}, \tag{8.7}$$

where C is concentration of the reactant (perovskite precursor), C_0 is its equilibrium concentration, k is the Boltzmann constant, T is the reaction temperature and V_m is the volume of a solute particle. Because crystallization occurs in solution phase, we should consider both ΔG and surface tension σ between surface of crystal and solution for perovskite crystallization kinetics. This relationship can be simply represented as [54–56]:

$$\Delta G_S = \overline{\sigma_s}\ , \tag{8.8}$$

where $\overline{\sigma_s}$ is the average of the surface tension of the crystal surface and of the solution (ΔG_S). Considering that perovskite crystal have cubic structures, Eq. (8.6) can be described as [54–56]:

$$\Delta \mathrm{G} = -a^3 \frac{kT}{V_\mathrm{m}} \ln\left(\frac{C}{C_0(T)}\right) + 6a^2\overline{\sigma_s}\ , \tag{8.9}$$

where a is the lattice parameter of perovskite cubic crystal. Differentiating (8.9) with respect to a yield

$$\frac{d\Delta G}{da} = -3a^2 \frac{kT}{V_\mathrm{m}} \ln\left(\frac{C}{C_0(T)}\right) + 12a\overline{\sigma_s} = 0\ . \tag{8.10}$$

Using Eq. (8.10), critical size a_c (8.11) and critical Gibbs free energy ΔG_c (8.12) at nucleation can be calculated as

$$a_c = \frac{4\overline{\sigma_s}}{\frac{kT}{V_\mathrm{m}} \ln\left((\frac{C}{C_0(T)}) \right)}\ . \tag{8.11}$$

$$\Delta G_c = \frac{32\overline{\sigma_S^2}}{\left(\left(\frac{kT}{V_\mathrm{m}}\right) \ln\left(\frac{C}{C_0(T)}\right)\right)^2}\ . \tag{8.12}$$

Gibbs free energy can change with crystal size. At perovskite crystal size > a_c, Gibbs free energy decreases, but at crystal size < a_c, Gibbs free energy increases. Thus, a perovskite crystal's growth mechanism is determined by (1) nucleation; perovskite nuclei > a_c and energy > ΔG_c grow spontaneously, and (2) space for crystal growth; perovskite crystals grow until they meet each other at grain boundaries [57, 58].

Average volume $V = a^3$ of the perovskite crystals is inversely proportional to the number of nuclei possible for nucleation and growth because a large number of nuclei decreases the growth space and thereby inhibits crystal growth (Fig. 8.4a,b) [57, 58]. According to statistical thermodynamics, the number of nuclei that have energy > ΔG_c is proportional to $e^{-\frac{\Delta G_c}{kT}}$. Thus, grain size A can be derived as

$$A \propto a \propto \sqrt[3]{\frac{1}{n}} \propto e^{\frac{\Delta G_c}{3kT}}\ . \tag{8.13}$$

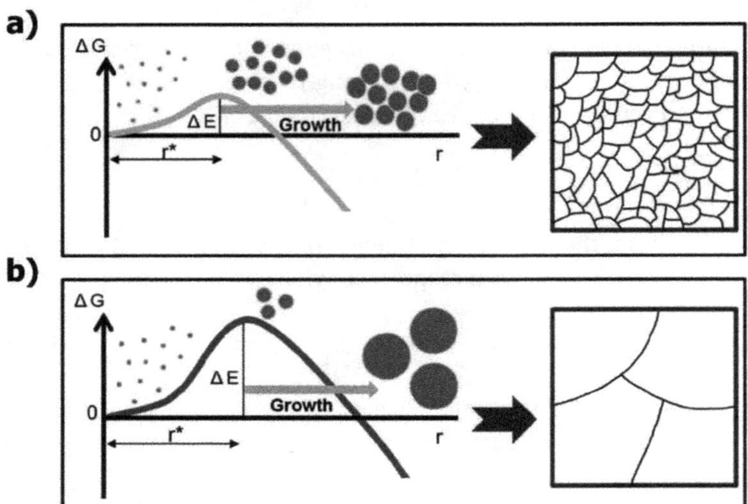

Figure 8.4 Schematics of crystal nucleation and growth according to the critical Gibbs free energy; (a) facilitated nucleation and prevented crystal growth, and (b) limited nucleation and promoted crystal growth. Reproduced with permission [58]. Copyright 2016, Royal Society of Chemistry.

The relation between *A* and solution concentration *X* can be obtained by combining Eq. (8.12) and (8.13):

$$\ln(A) = \frac{32\overline{\sigma_s^3}}{3kT\left(\frac{kT}{V_m}(\ln X - \ln C_0(T))\right)^2} + C' \tag{8.14}$$

We can experimentally fit this Eq. (8.14) in the one-step solution process and a two-step solution process, separately (Fig. 8.5a).

In the one-step method, Eq. (8.14) can be fitted to Eq. (8.15) when the $MAPbI_3$ crystal is fabricated in one step using chlorobenzene dripping, assuming that equilibrium solubility of perovskite is 0.403 M at 100°C [55, 56]:

$$\ln(A) = -\frac{0.742}{\left[\ln X - \ln C_0(T)\right]^2} + 5.992. \tag{8.15}$$

Equation 8.15 indicates that grain size increases as solution concentration increases (supersaturation concentration) (Fig. 8.5b), and that crystal growth rather than nucleation is the dominant mechanism in the one-step crystallization process. This conclusion is reached because in the one-step solution process, especially when the solvent dripping process is used, additional nucleation sites cannot be formed after initial nucleation stages, so growth of nuclei is the only possible crystallization mechanism [56].

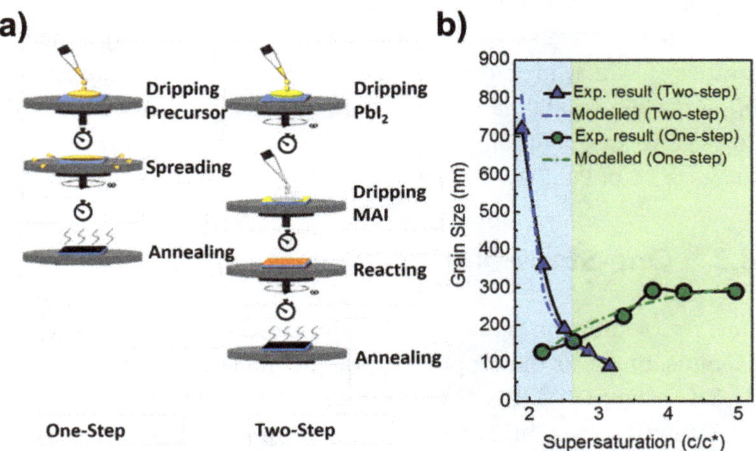

Figure 8.5 (a) Schematics of perovskite PC film deposition processes (one-step process and two-step process), (b) grain sizes according to the supersaturation concentration in one-step process and two-step process (here, c* is the saturation concentration of perovskite precursor). Reproduced with permission [56]. Copyright 2017, American Chemical Society.

In the two-step method, crystallization occurs during the 2nd step in which *AX* is deposited on top of an inorganic MX_2 layer:

$$MX_2\ (s) + AX\ (sol) \leftrightarrow AMX_3(s), \tag{8.16}$$

The relation between *A* and *X* during formation of $MAPbI_3$ crystals can be experimentally fitted as [55, 56]:

$$\ln(A) = \frac{1.22}{[\ln X - \ln c_0(T)]^2} + 3.73. \tag{8.17}$$

According to Eq. (8.17), grain size decreases as *X* of AX is increased [55, 56]. This equation corresponds to the relationship between grains size and precursor concentration in the two-step solution process: as precursor concentration increases, nucleation density increases and the space for nuclei to grow decreases, so grain size is small (Fig. 8.5b) [59]. Based on this observation, we speculate that nucleation is the dominant mechanism during the two-step method, because small grain must be a result of increased density of nuclei as precursor concentration increases [59].

Because PLQY of perovskite PC films tends to increase with decreasing perovskite grain size due to higher probability of excitonic radiative recombination in confined system [4], different strategies which promote the nucleation and prevent the growth of perovskite crystal have been tried to reduce the perovskite grain size in one-step solution process and two-step solution process, and increase LE of PeLEDs.

8.3 One-Step Solution Process

One-step solution process is the simplest method to form perovskite PC films, thus was first applied for the demonstration of PeLEDs [2, 3]. The bright OIP LEDs at RT exceeding 100 cd/m^2 were reported by Friend's group in 2014 [2]. They demonstrated NIR-emitting inverted LEDs with EQE ~ 0.76 % and radiance ~ 13.2 W sr^{-1} m^{-2} and green-emitting inverted OIP LEDs with EQE ~ 0.1%, CE ~ 0.3 cd/A, and luminance L ~ 364 cd/m^2 [2]. Three months later, Lee's group achieved bright green-emitting OIP LEDs with EQE ~ 0.125%, CE ~ 0.577 cd/A, and L ~ 417 cd/m^2 by managing exciton quenching at the interface between perovskite emission layer and hole injection layer (HIL) using a fluorinated conducting polymer compositions [3]. These initial PeLEDs showed relatively low device efficiencies because OIP PCs have many defects and pinholes, and rough surfaces. Thus, after these papers, researchers have tried to improve the LE of PeLEDs by various methods such as modification of the perovskite emission layer [4, 60], management of charge-carrier quenching inside the perovskite grains and at the grain boundary [4], defect passivation [61], fabrication of uniform perovskite layer and control of their compositions [4, 26, 62–64]. In this section, we specifically describe crystal growth and nucleation mechanism in one-step solution process, various fabrication methods of uniform OIP and

IHP films, and applications on PeLEDs according to the perovskite film formation methods that were used.

8.3.1 Crystal Formation Mechanism

AX precursor and PbX_2 precursor were mixed together in polar solvent with high dipole moment (e.g., N,N-dimethylformamide (DMF), dimethylsulfoxide (DMSO), gbutyrolactone (GBL)). Then perovskite PC films were formed by spincoating the precursor solutions on a substrate. During the spincoating process, the solvent is spun off by centrifugal force and thus the perovskite solution on the substrate are driven to be saturated gradually (Fig. 8.6). As the concentration of perovskite solution increases to greater than supersaturation concentration, PbX_2 particles which have lower solubility in polar solvent than does MAX tend to crystallize first, then react with MAX; these processes correspond to the nucleation of crystal nuclei [65, 66]. Then perovskite quasi-film gets gelled and becomes visibly hazy due to growing perovskite crystal. During continuous spincoating, perovskite crystals gradually grow as the polar solvent is ejected. After all the remaining solvent has evaporated, they stop growing [65].

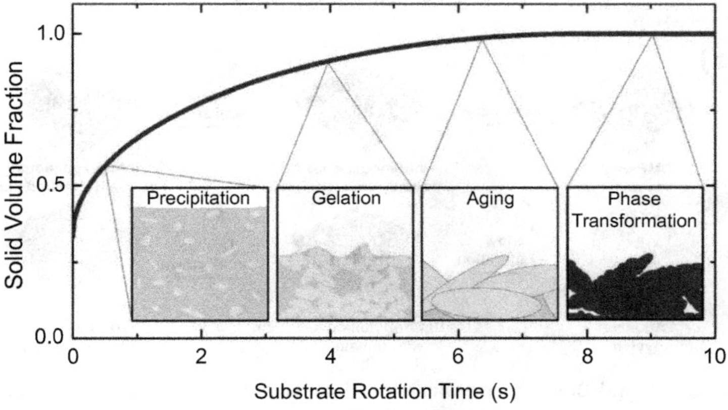

Figure 8.6 Schematics of perovskite crystal formation in one-step solution process. Reproduced with permission [65]. Copyright 2016, Royal Society of Chemistry.

8.3.2 Retarded Crystallization by Adding Additives

Crystal size and growth dynamics in the one-step solution process are dominantly affected by the supersaturation concentration (8.18) [67]:

$$\varphi = k_g(C_v^i - C_v^0) \tag{8.18}$$

where φ is evaporation rate (m/s), C_v^i is the concentration of solution at the interface, C_v^o is the concentration of solvent in the vapor phase, and k_g is the mass transfer coefficient. This equation reveals that the growth rate and size of crystals can be controlled by tuning the solubility of perovskite precursors and the supersaturation concentration [67].

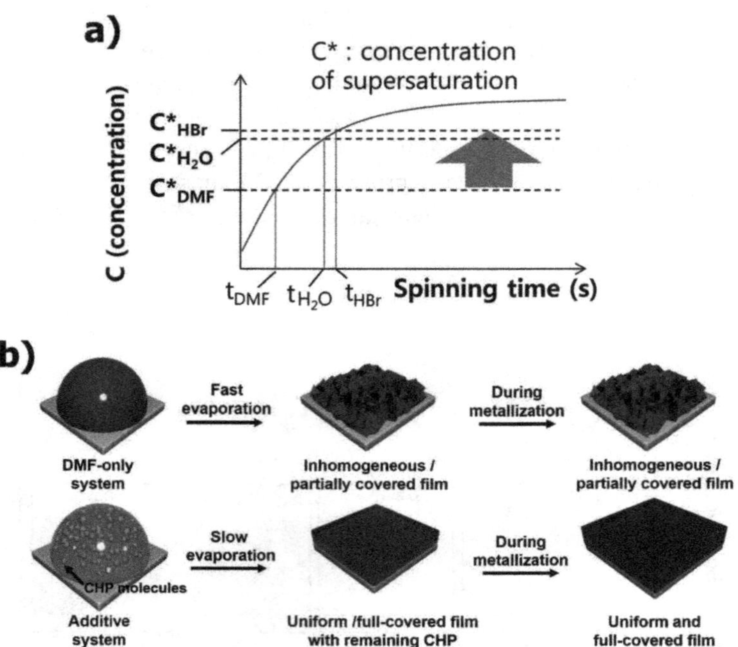

Figure 8.7 (a) Delayed crystallization time of perovskite solution by adding water and HBr additives. Reproduced with permission from Ref. [67]. Copyright 2014, WILEY-VCH Verlag GmbH & Co. KGaA, Weinheim. (b) Schematic illustrations of crystallization process and perovskite PC films by adding organic additives with low vapor-pressure. Reproduced with permission from Ref. [68]. Copyright 2015, Nature Publishing Group.

Formation of perovskite crystal by the one-step solution process yields large, sparsely-coated cubic crystals (> micrometers) [56]; which cause a rough surface and many pinholes, and thereby limit the LE of PeLEDs [4]. The rough surface of these perovskite PC films can be dense and smoothened by adding additives: (1) acid additives such as HBr increase the viscosity, solubility and supersaturation concentration of the perovskite solutions (Fig. 8.7a); (2) organic additives (e.g., N-cyclohexyl-2-pyrrolidone (CHP)) with low vapor pressure reduce solvent evaporation time (Fig. 8.7b) [69–71]. These methods retard crystallization and thereby increase the density of perovskite PC films and improve their film coverages. PeLEDs with HBr acid additives showed EQE ~ 0.1%, CE ~ 0.43 cd/A and $L \sim 3{,}490\ \text{cd/m}^2$ [69].

Although these additive methods improve surface morphology and film density, increased grain size induced by delayed crystallization time can facilitate the charge separation and limit LE. Prohibiting crystal growth during the initial growth stages reduces the grain size of perovskite crystal and prevents the separation of electron and hole carriers; therefore, prevention of this early growth can be an effective method to increase PLQY of perovskite PC films and LE of PeLEDs.

8.3.3 Prevented Crystal Growth by Inhibitors

Crystal growth dynamics during the one-step solution process can be affected by impurity molecules (here, we call inhibitors). Inhibitors are poorly-soluble in solution and can be adsorbed on growing crystal 2D terraces, thus crystallization mechanisms follow the Kossel-Stranski terrace-step-kink model [72] and crystallization rate can be described as

$$w(C_I) \sim \begin{cases} w(0)e^{-K_I C_I}, & C_I < C_{CI} \\ 0, & C_I > C_{CI} \end{cases} \qquad K_I = \frac{4Kf^{\frac{1}{2}}\gamma V_m}{kT\sigma\pi^{\frac{1}{2}}}, \tag{8.19}$$

where w [m s^{-1}] is crystallization rate, C_I [m s^{-3}] is concentration of inhibitors, C_{CI} is a critical value of C_I, K_I is the constant of inhibitor concentration effect on the crystallization rate, K is the absorption constant, f [m^2] is area partially occupied by the adsorbed inhibitors

$[m^2]$, γ[J m^{-2}] is growth step energy, V_m [m^3] is unit volume of crystal structure, T is the absolute temperature and σ [dimensionless] is relative solution supersaturation. w decreases exponentially as C_I increases, and converges to zero when $C_I > C_{CI}$. When C_I is low, crystallization rate becomes

$$\lim_{C\to 0} w(C_I) = (1 - K_I C_I) w(0), \tag{8.20}$$

which means that w decreases linearly as C_I increases. This analysis reveals that the presence of inhibitors prevents growth of perovskite crystal and reduces grain size.

To exploit this phenomenon, various methods such as (i) mixing organic molecules, (ii) mixing organic ammonium, and (iii) controlling the precursor ratio in perovskite precursor solutions have been used to reduce the perovskite grain size and fabricate uniform perovskite PC films. Increasing the Gibbs free energy for perovskite growth also induces relatively faster nucleation rate than growth rates, and therefore yields uniform perovskite PC films with small grains (Fig. 8.4) [58].

8.3.3.1 Inhibited crystal growth by organic molecules

Large amounts of organic additives such as small organic molecules and polymers can inhibit growth of perovskite crystals. Large amounts of (4,4-bis(N-carbazolyl)-1,1-biphenyl (CBP)) in $MAPbBr_3$/DMF solutions prevent the crystal growth and induce uniform CBP:perovskite composite films with small perovskite crystal grains [73]. As the CBP weight ratio compared to $MAPbBr_3$ precursors increased from 1:1 to 15:1, the average size of $MAPbBr_3$ grain decreased from ~76 nm to ~10 nm; as a result, PL was blue-shifted due to quantum confinement effects.

Polymers such as poly(ethylene oxide) (PEO) [29, 30], polyvinylcarbazole (PVK) [74] and polyimide precursor dielectric (PIP) [75] can be mixed with $MAPbBr_3$ or $CsPbBr_3$ precursors to inhibit crystal growth. These inhibitors efficiently prevent crystal growth and induce uniform polymer:perovskite composite films with perovskite grain ~ 100 nm and surface coverage > 95% (Fig. 8.8a). With these methods, efficient PeLEDs were demonstrated (EQE ~ 1.1%, CE ~ 4.91 cd/A, L ~ 21,014 cd/m^2 for PeLEDs based on PEO:$MAPbBr_3$ (Fig. 8.8b–e) [29, 30]; EQE ~ 0.64%, CE ~ 2.37 cd/A,

L ~ 1,561 cd/m^2 for PeLEDs based on PIP:$MAPbBr_3$ [75]; EQE ~ 4.26%, CE ~ 15.67 cd/A for PeLEDs based on PEO:$CsPbBr_3$ [27,76]; EQE ~ 5.7 %, power efficiency PE ~ 14.1 lm/W for PeLEDs based on PEO:PVP:$CsPbBr_3$ [77]).

Basic salts such as NABr can also increase the number of perovskite crystals and surface coverage [78]. NaBr may also passivate bromide vacancies and thereby reduce the number of ionic defects and improve LE in PeLEDs (EQE ~ 0.17%, CE ~ 0.72 cd/A) compared to pristine PeLEDs (EQE ~ 0.02%, CE ~ 0.1 cd/A).

8.3.3.2 Hindered crystal growth by mixed organic ammonium

Mixing large organic ammonium molecules in $MAPbX_3$ precursor solution also hinders crystal growth and thus, induces small crystal grain size and excellent surface coverage [25, 26, 62, 63, 79–81]. Organic ammonium-mixed OIP PC films were formed with quasi-2D structures, which efficiently confined charge carriers, increased E_B, induced energy transfer of electron-hole pairs to the smallest bandgap, and reduced trap density and nonradiative trap-related recombination [64, 82]. These effects helped to achieve high PLQY (34%), which was much higher than that of pure $MAPbBr_3$ PC films (<2–3%) [62], and also helped to achieve highly efficient PeLEDs (CE ~ 4.9 cd/A, L ~ 2,935 cd/m^2 for green-emitting PeLEDs based on $PEA_2MA_{m-1}Pb_mBr_{3m+1}$ (here, PEA is phenylethyl ammonium) [62]; EQE ~ 10.4% for green-emitting PeLEDs based on $BA_{0.17}MA_{0.83}PbBr_3$ and EQE ~ 9.3% for NIR-emitting PeLEDs based on $BA_{0.17}MA_{0.83}PbI_3$ (here, BA is *n*-butylammonium) [25]; EQE ~ 11.7% for NIR-emitting PeLEDs based on $(NMA)_2(FAPbI_3)PbI_4$ (here, NMA is 1-naphthylmethylamine) [26].

8.3.3.3 Restricted crystal growth by precursor ratio control

Equimolar MABr:$PbBr_2$ ratio (1:1) induced non-uniform morphology with large cubic crystals (> micrometers), incomplete reaction between MABr and $PbBr_2$, and metallic Pb atoms because some Br atoms were lost [4] or PbX_2 tended to aggregate and precipitate earlier than AX [66]. Therefore, control of the precursor ratio for perovskite (ABX_3) can also effectively improve the surface morphology and LE of the perovskite PC films.

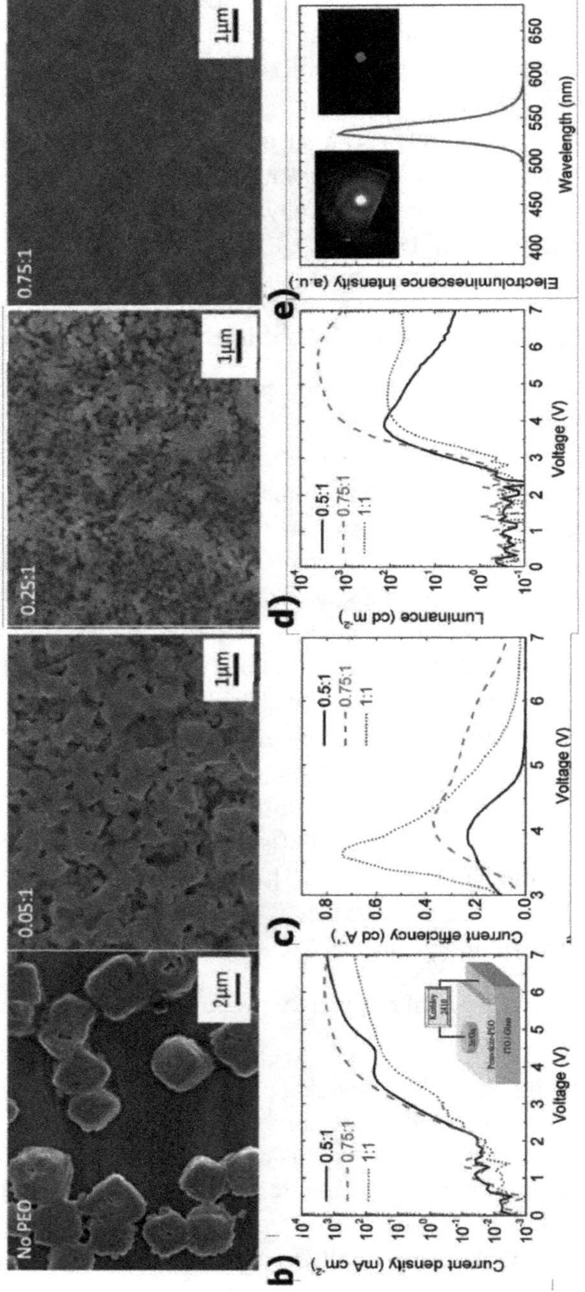

Figure 8.8 (a) SEM images of PEO:$MAPbBr_3$ composite films with different PEO:$MAPbBr_3$ ratios, (b) current density characteristics, (c) current efficiency characteristics, (d) luminance characteristics, and (e) EL spectrum and emitting images of PeLEDs based on PEO:$MAPbBr_3$ composite films. Reproduced with permission from Ref. [29]. Copyright 2015, WILEY-VCH Verlag GmbH & Co. KGaA, Weinheim.

In $MAPbBr_3$, as the molar ratio of MABr increased to 2.2:1, morphology of $MAPbBr_3$ PC films improved and perovskite grain size decreased because excess MABr prevented crystallization of perovskite by inducing dewetting of perovskite grains on the underlayer [66]. With optimal MABr:$PbBr_2$ ratio (2.2:1), high-efficiency $MAPbBr_3$ PeLEDs with EQE ~ 3.38%, CE ~ 15.26 cd/A and L ~ 6,124 cd/m^2 were achieved; these were much higher than in PeLEDs with equimolar MABr:$PbBr_2$ ratio (EQE ~ 0.004%, CE ~ 0.02 cd/A and L ~ 28.8 cd/m^2).

When the ratio of MABr to $PbBr_2$ is slightly high (1.05:1 in MAPb:$PbBr_2$), metallic Pb atoms are removed from the resulting $MAPbBr_3$ PC films with equimolar ratio of MABr:$PbBr_2$ [4]. This method achieved high CE of 21.4 cd/A in PeLEDs based on $MAPbBr_3$ film fabricated by solvent-based nanocrystal pinning (NCP) process, which is much higher than did PeLEDs with MABr:$PbBr_2$ = 1:1 (CE ~ 0.183 cd/A). Control of perovskite precursor ratio can also be effective to improve the LE of IHP PC films and IHP PeLEDs [83]. With this method, PL lifetime of $CsPbBr_3$ PC films increased from 5.06 ns to 45.7 ns and their PLQY increased from 0.5% to 33.6% due to passivation of surface traps and reduced nonradiative recombination in the $CsPbBr_3$-CsBr (1:0.4mol) film. This use of excess CsBr also achieved high EL efficiencies (CE ~ 0.57 cd/A, L ~ 7,276 cd/m^2 and high operating stability in air (L > 100 cd/m^2 for > 15 h).

Halide anion compositions in perovskite structures also affect the surface morphology [84, 85]. OIP PC films with mixed halides ($CH_3NH_3PbI_xBr_xCl_{3-x-y}$) induced smaller crystal (> 100 nm) than did pure $MAPbBr_3$ crystal (> micrometers) although both films were fabricated by simple spin-coating [85].

8.3.4 Fast Termination of Crystal Growth

NCP is a simple but effective process to fabricate uniform OIP PC films with small grain sizes. Dripping volatile non-polar solvents (e.g., chloroform) on the spinning wet perovskite solution during spin-coating process can wash away the good solvents (e.g., DMF, DMSO and GBL), induce immediate termination of crystal growth and thereby decrease grain size (Fig. 8.9a–d) [4, 74, 86]. With these NCP methods using CF dripping solvent, uniform $MAPbBr_3$ films with small sized nano-grain (100–250 nm) were achieved. These grain size can be further reduced to 50–150 nm by adding small molecular organic additives (inhibitors) ((2,2′,2″-(1,3,5-benzinetriyl)-tris(1-phenyl-1-H-benzimidazole) (TPBI)) in CF (this process was named additive-based NCP (A-NCP)) because additives can inhibit crystal growth (Fig. 8.9e). This NCP process can be differentiated from other antisolvent dripping methods, which have purpose to achieve uniform large grain size of perovskites to minimize hysteresis and structural defects in perovskite solar cells, and therefore do not use additives inhibiting crystal growth [1, 87–89]. PeLEDs using A-NCP process showed bright PC films and significantly high LE (EQE ~ 8.52% and CE ~ 42.9 cd/A) (Fig. 8.9f,g) [4].

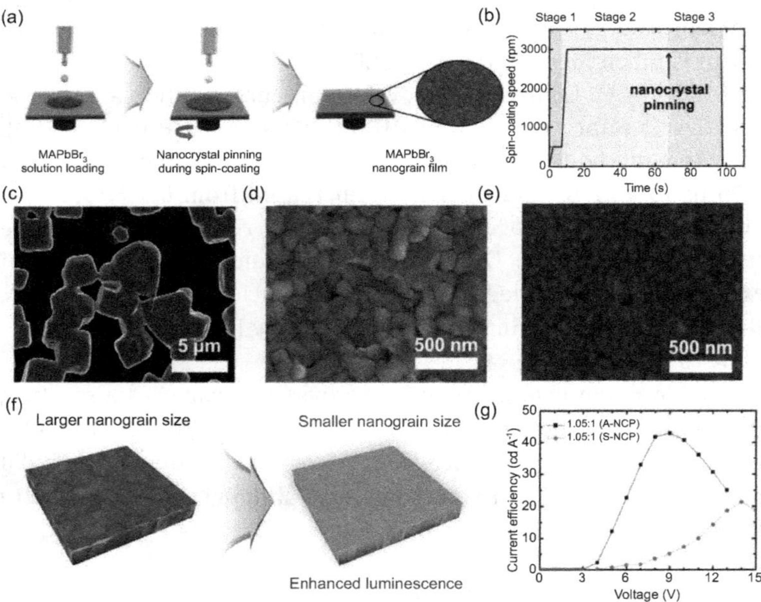

Figure 8.9 (a) Schematics of nanocrystal pinning (NCP) process; (b) $MAPbBr_3$ crystal formation profile with NCP process according to the spin-coating time; SEM image of (c) pristine $MAPbBr_3$ film, (d) $MAPbBr_3$ film with NCP process, and (e) $MAPbBr_3$ film with A-NCP process; (f) illustrations of $MAPbBr_3$ film with different grain sizes; and (g) CE characteristics of PeLEDs based on NCP and A-NCP process. Reproduced with permission from Ref. [4], Copyright 2015, American Association for the Advancement of Science.

8.3.5 Facilitated Crystal Nucleation by Interfacial Treatments

The nucleation process has also an important influence on morphology of perovskite PC films. Nucleation processes are divided into two types: homogeneous nucleation and heterogeneous nucleation. Homogeneous nucleation occurs within the bulk solution; heterogeneous nucleation takes place on foreign nuclei or surfaces. According to classical crystallization theory, the energy barrier is lower for heterogeneous nucleation than for homogeneous nucleation [90]. This difference implies that facilitation of heterogeneous nucleation of perovskite crystals on the surface of underlayers, will yield numerous nuclei and yield uniform perovskite PC films with small grains.

Assuming that nucleation and growth of perovskite crystal occur in solution phase, Gibbs free energy $\Delta G_{\text{heterogeneous}}$ for heterogeneous nucleation and Gibbs free energy $\Delta G_{\text{homogeneous}}$ for homogeneous nucleation are expressed as (8.21, 8.22),

$$\Delta G_{\text{heterogeneous}} = \Delta G_{\text{homogeneous}} \times f(\theta) \tag{8.21}$$

$$f(\theta) = \frac{2 - 3\cos\theta + cos^3\theta}{4}, \tag{8.22}$$

where θ is the contact angle of the liquid on a solid surface [91]. These equations imply that as the wettability of perovskite solution on the under-layer increases (θ decreases), $\Delta G_{\text{heterogeneous}}$ decreases and heterogeneous nucleation is facilitated; this change implies increase in the number of nuclei that can be formed on top of under-layer and decrease in grain size. In contrast, as the wettability of perovskite solution decreases and $\Delta G_{\text{heterogeneous}}$ increases, the number of nuclei decreases, and grain size increases [92]. Thus, many researchers have tried to facilitate heterogeneous nucleation by increasing the wettability of solution on substrates to achieve uniform and bright perovskite PC films with small grain size [57, 93–96].

A hydrophilic NiO_x interlayer, amino acid self-assembled monolayers (SAMs), and branched polyethyleneimine (PEI) interlayer on ZnO all increased the wettability of perovskite/polar solvent solution by improving the hydrophilicity of underlayers.

These surface treatments enabled formation of uniform perovskite PC films with good film surface coverage, small grain size and high crystallinity [95–97]. Furthermore, these interlayer can also increase LE of PeLEDs by facilitating charge injection into perovskite emission layer and preventing quenching of charge carriers at the interfaces [3, 95]. With these surface-treatment processes, PeLEDs with high LE were achieved (CE ~ 15.9 cd/A, L > 65,300 cd/m^2 for PeLEDs based on $MAPbBr_3$ that had a NiO_x interlayer [97]).

8.4 Two-Step Solution Process

The two-step solution process can also be effective to fabricate uniform OIP PC films, because nucleation is the main mechanism in this fabrication method, rather than crystal growth as described in the Section 8.2.3 (Fig. 8.10a) [98]. The crystal size and PLQY of OIP PC films obtained using two-step solution processing were controlled by (i) concentration of AX precursor solution and BX_2 precursor solution [99] and (ii) loading time of the second-deposited precursor solution [100]. Increasing concentration of MAX precursor solution increased the number of perovskite nuclei and reduced the final crystal size (Fig. 8.10b,c) [59, 99]. This method can facilitate sufficient conversion from PbX_2 to $MAPbX_3$ crystals and increase the surface coverage [99, 100].

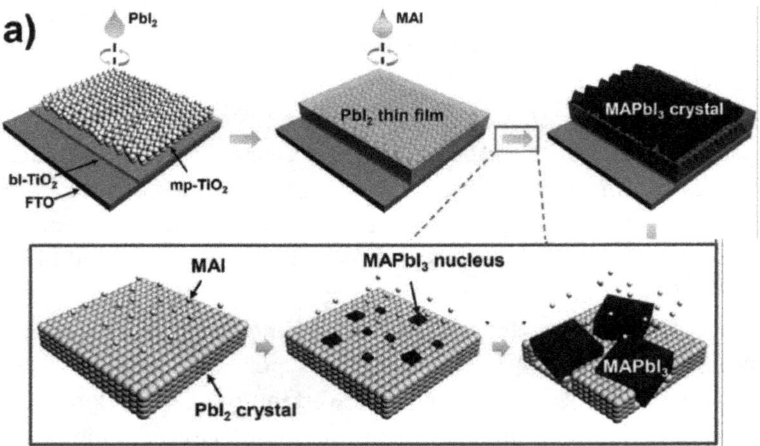

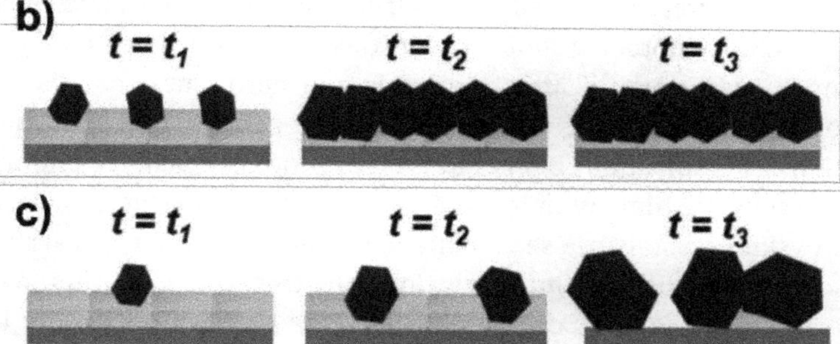

Figure 8.10 (a) Schematic illustrations of fabrication and crystallization mechanism for $MAPbI_3$ PC films using two-step solution process. Reproduced with permission from Ref. [55]. Copyright 2015, The Royal Society of Chemistry. Schematic figures of crystal nucleation and growth using two-step solution process with (b) high MAX solution concentration and (c) low MAX solution concentration. Reproduced with permission from Ref. [59]. Copyright 2017, American Chemical Society.

$FAPbBr_3$ PC films fabricated by spin-coating FABr/isopropyl alcohol solution on top of $PbBr_2$ showed uniform surface morphology with root mean-squared error < 20 nm and small average grain size of 100 - 200 nm [98]. $MAPbBr_3$ films obtained using the two-step process with optimum conditions showed the highest PLQY of 24%, and obtained EL efficiencies of EQE ~ 0.023% and CE ~ 0.1 cd/A [99].

8.5 Conclusion

OIPs and IHPs have high color purity (FWHM < 20 nm) that is independent of grain size and thus, they have been expected to be used as vivid natural color emitters in the near future. However, perovskite PC films undergo dissociation and nonradiative recombination of electron-hole pairs, and have rough surface morphology and numerous surface traps that limit the PLQY of perovskite PC films and the LE of PeLEDs. Thus, to increase the PLQY and LE of perovskite PC films and PeLEDs, (i) dimension of perovskite grain should be reduced to prevent the charge carrier separation and to increase the radiative recombination rate of charge carriers, and (ii) uniform perovskite PC films should be fabricated to eliminate leakage current in PeLEDs. To solve these problems and to increase the LE of PeLEDs, several strategies have been developed.

This chapter introduced device structure and working mechanism of PeLEDs, optoelectronic properties of perovskite emitters, fundamental crystallization mechanism of perovskite crystals and methods to modify perovskite PC films to achieve high-efficiency PeLEDs. Modifications of crystal growth processes achieve uniform perovskite PC films with small grains by preventing the growth of perovskite crystals; these can spatially confine the charge carriers and prevent them from separating into free carriers. Interfacial treatments in PeLEDs can help to fabricate uniform perovskite PC films with small grain size by increasing the crystal nucleation rate. Two-step solution process can also achieve the uniform perovskite PC films with small grain because crystal nucleation is main crystallization mechanism rather than crystal growth. With these research strategies and the exertion of researchers from diverse disciplines, LE of PeLEDs been increased dramatically. These great achievements in LE of PeLEDs stimulate further growth of perovskite emitters in scientific fields and diverse applications of them in future display and solid-state lightings.

Acknowledgments

This work was supported by the National Research Foundation of Korea (NRF) grant funded by the Korea government (Ministry of Science, ICT & Future Planning) (Grant No. NRF-2016R1A3B1908431).

References

1. Kim, Y.-H., Cho, H., and Lee, T.-W. (2016). Metal halide perovskite light emitters. *Proc. Natl. Acad. Sci. U. S. A.*, **113**, 11694–11702.
2. Tan, Z.-K., Moghaddam, R. S., Lai, M. L., Docampo, P., Higler, R., Deschler, F., Price, M., Sadhanala, A., Pazos, L. M., Credgington, D., Hanusch, F., Bein, T., Snaith, H. J., and Friend, R. H. (2014). Bright light-emitting diodes based on organometal halide perovskite. *Nat. Nanotechnol.*, **9**, 687–692.
3. Kim, Y.-H., Cho, H., Heo, J. H., Kim, T.-S., Myoung, N., Lee, C.-L., Im, S. H., and Lee, T.-W. (2015). Multicolored organic/inorganic hybrid perovskite light-emitting diodes. *Adv. Mater.*, **27**, 1248–1254.

4. Cho, H., Jeong, S.-H., Park, M.-H., Kim, Y.-H., Wolf, C., Lee, C.-L., Heo, J. H., Sadhanala, A., Myoung, N., Yoo, S., Im, S. H., Friend, R. H., and Lee, T.-W. (2015). Overcoming the electroluminescence efficiency limitations of perovskite light-emitting diodes. *Science*, **350,** 1222–1225.

5. Hattori, T., Taira, T., Era, M., Tsutsui, T., and Saito, S. (1996). Highly efficient electroluminescence from a heterostructure device combined with emissive layered-perovskite and an electron-transporting organic compound. *Chem. Phys. Lett.*, **254**, 103–108.

6. Chondroudis, K., and Mitzi, D. B. (1999). Electroluminescence from an organic–inorganic perovskite incorporating a quaterthiophene dye within lead halide perovskite layers. *Chem. Mater.*, **11**, 3028–3030.

7. Koutselas, I., Bampoulis, P., Maratou, E., Evagelinou, T., Pagona, G., and Papavassiliou, G. C. (2011). Some unconventional organic-inorganic hybrid low-dimensional semiconductors and related light-emitting devices. *J. Phys. Chem. C*, **115**, 8475–8483.

8. Tang, C. W., Vanslyke, S. A., and Chen, C. H. (1989). Electroluminescence of doped organic thin films. *J. Appl. Phys.*, **65**, 3610–3616.

9. Adachi, C., Baldo, M. A., Thompson, M. E., and Forrest, S. R. (2001). Nearly 100% internal phosphorescence efficiency in an organic light emitting device. *J. Appl. Phys.*, **90**, 5048–5051.

10. Baldo, M. A., O 'brien, D. F., You, Y., Shoustikov, A., Sibley, S., Thompson, M. E., and Forrest, S. R. (1998). Highly efficient phosphorescent emission from organic electroluminescent devices. *Nature*, **395**, 151–154.

11. Han, T.-H., Choi, M.-R., Woo, S.-H., Min, S.-Y., Lee, C.-L., and Lee, T.-W. (2012). Molecularly controlled interfacial layer strategy toward highly efficient simple-structured organic light-emitting diodes. *Adv. Mater.*, **24**, 1487–1493.

12. Uoyama, H., Goushi, K., Shizu, K., Nomura, H., and Adachi, C. (2012). Highly efficient organic light-emitting diodes from delayed fluorescence. *Nature*, **492**, 234–238.

13. Kim, Y.-H., Wolf, C., Cho, H., Jeong, S.-H., and Lee, T.-W. (2016). Highly efficient, simplified, solution-processed thermally activated delayed-fluorescence organic light-emitting diodes. *Adv. Mater.*, **28**, 734–741.

14. Colvin, V. L., Schlamp, M. C., and Alivisatos, A. P. (1994). Light-emitting diodes made from cadmium selenide nanocrystals and a semiconducting polymer. *Nature*, **370**, 354–357.

15. Dabbousi, B. O., Bawendi, M. G., Onitsuka, O., and Rubner, M. F. (1995). Electroluminescence from CdSe quantum-dot/polymer composites. *Appl. Phys. Lett.*, **66**, 1316–1318.

16. Schlamp, M. C., Peng, X., and Alivisatos, A. P. (1997). Improved efficiencies in light emitting diodes made with CdSe(CdS) core/shell type nanocrystals and a semiconducting polymer. *J. Appl. Phys.*, **82**, 5837–5842.

17. Coe, S., Woo, W.-K., Bawendi, M., and Bulović, V. (2002). Electroluminescence from single monolayers of nanocrystals in molecular organic devices. *Nature*, **420**, 800–802.

18. Cho, K.-S., Lee, E. K., Joo, W.-J., Jang, E., Kim, T.-H., Lee, S. J., Kwon, S.-J., Han, J. Y., Kim, B.-K., Choi, B. L., and Kim, J. M. (2009). High-performance crosslinked colloidal quantum-dot light-emitting diodes. *Nat. Photonics*, **3**, 341–345.

19. Mashford, B. S., Stevenson, M., Popovic, Z., Hamilton, C., Zhou, Z., Breen, C., Steckel, J., Bulovic, V., Bawendi, M., Coe-Sullivan, S., and Kazlas, P. T. (2013). High-efficiency quantum-dot light-emitting devices with enhanced charge injection. *Nat. Photonics*, **7**, 407–412.

20. Dai, X., Zhang, Z., Jin, Y., Niu, Y., Cao, H., Liang, X., Chen, L., Wang, J., and Peng, X. (2014). Solution-processed, high-performance light-emitting diodes based on quantum dots. *Nature*, **515**, 96–99.

21. Kim, H.-S., Lee, C.-R., Im, J.-H., Lee, K.-B., Moehl, T., Marchioro, A., Moon, S.-J., Humphry-Baker, R., Yum, J.-H., Moser, J. E., Grätzel, M., and Park, N.-G. (2012). Lead iodide perovskite sensitized all-solid-state submicron thin film mesoscopic solar cell with efficiency exceeding 9%. *Sci. Rep.*, **2**, 591.

22. Stranks, S. D., Eperon, G. E., Grancini, G., Menelaou, C., Alcocer, M. J. P., Leijtens, T., Herz, L. M., Petrozza, A., and Snaith, H. J. (2014). Electron-hole diffusion lengths exceeding 1 micrometer in an organometal trihalide perovskite absorber. *Science*, **342**, 341–344.

23. Yang, W. S., Noh, J. H., Jeon, N. J., Kim, Y. C., Ryu, S., Seo, J., and Seok, S. Il. (2015). High-performance photovoltaic perovskite layers fabricated through intramolecular exchange. *Science*, **348**, 1234–1237.

24. Jeon, N. J., Noh, J. H., Yang, W. S., Kim, Y. C., Ryu, S., Seo, J., and Seok, S. Il. (2014). Compositional engineering of perovskite materials for high-performance solar cells. *Nature*, **517**, 476–480.

25. Xiao, Z., Kerner, R. A., Zhao, L., Tran, N. L., Lee, K. M., Koh, T.-W., Scholes, G. D., and Rand, B. P. (2017). Efficient perovskite light-emitting diodes featuring nanometre-sized crystallites. *Nat. Photonics*, **11**, 108–115.

26. Wang, N., Cheng, L., Ge, R., Zhang, S., Miao, Y., Zou, W., Yi, C., Sun, Y., Cao, Y., Yang, R., Wei, Y., Guo, Q., Ke, Y., Yu, M., Jin, Y., Liu, Y., Ding, Q., Di, D., Yang, L., Xing, G., Tian, H., Jin, C., Gao, F., Friend, R. H., Wang, J., and

Huang, W. (2016). Perovskite light-emitting diodes based on solution-processed self-organized multiple quantum wells. *Nat. Photonics*, **10**, 699–704.

27. Ling, Y., Tian, Y., Wang, X., Wang, J. C., Knox, J. M., Perez-Orive, F., Du, Y., Tan, L., Hanson, K., Ma, B., and Gao, H. (2016). Enhanced optical and electrical properties of polymer-assisted all-inorganic perovskites for light-emitting diodes. *Adv. Mater.*, **28**, 8983–8989.

28. Zhang, L., Yang, X., Jiang, Q., Wang, P., Yin, Z., Zhang, X., Tan, H., Yang, Y. M., Wei, M., Sutherland, B. R., Sargent, E. H., and You, J. (2017). Ultra-bright and highly efficient inorganic based perovskite light-emitting diodes. *Nat. Commun.*, **8**, 15640.

29. Li, J., Bade, S. G. R., Shan, X., and Yu, Z. (2015). Single-layer light-emitting diodes using organometal halide perovskite/poly(ethylene oxide) composite thin films. *Adv. Mater.*, **27**, 5196–5202.

30. Bade, S. G. R., Li, J., Shan, X., Ling, Y., Tian, Y., Dilbeck, T., Besara, T., Geske, T., Gao, H., Ma, B., Hanson, K., Siegrist, T., Xu, C., and Yu, Z. (2016). Fully printed halide perovskite light-emitting diodes with silver nanowire electrodes. *ACS Nano*, **10**, 1795–1801.

31. Saba, M., Cadelano, M., Marongiu, D., Chen, F., Sarritzu, V., Sestu, N., Figus, C., Aresti, M., Piras, R., Geddo Lehmann, A., Cannas, C., Musinu, A., Quochi, F., Mura, A., and Bongiovanni, G. (2014). Correlated electron-hole plasma in organometal perovskites. *Nat. Commun.*, **5**, 5049.

32. Kim, Y.-H., Wolf, C., Kim, Y.-T., Cho, H., Kwon, W., Sadhanala, A., Park, C. G., Rhee, S.-W., Im, S. H., Friend, H. R., and Lee, T.-W. (2017). Highly efficient light-emitting diodes of colloidal metal-halide perovskite nanocrystals beyond quantum size. *ACS Nano*, **11**, 6586–6593.

33. Dvorak, M., Wei, S. H., and Wu, Z. (2013). Origin of the variation of exciton binding energy in semiconductors. *Phys. Rev. Lett.*, **110**, 1–5.

34. Galkowski, K., Mitioglu, A., Miyata, A., Plochocka, P., Portugall, O., Eperon, G. E., Wang, J. T.-W., Stergiopoulos, T., Stranks, S. D., Snaith, H. J., and Nicholas, R. J. (2016). Determination of the exciton binding energy and effective masses for methylammonium and formamidinium lead tri-halide perovskite semiconductors. *Energy Environ. Sci.*, **9**, 962–970.

35. Sestu, N., Cadelano, M., Sarritzu, V., Chen, F., Marongiu, D., Piras, R., Mainas, M., Quochi, F., Saba, M., Mura, A., and Bongiovanni, G. (2015). Absorption f-sum rule for the exciton binding energy in methylammonium lead halide perovskites. *J. Phys. Chem. Lett.*, **6**, 4566–4572.

36. Wright, A. D., Verdi, C., Milot, R. L., Eperon, G. E., Pérez-Osorio, M. A., Snaith, H. J., Giustino, F., Johnston, M. B., and Herz, L. M. (2016).

Electron–phonon coupling in hybrid lead halide perovskites. *Nat. Commun.*, **7**.

37. Kumar, A., Kumawat, N. K., Maheshwari, P., and Kabra, D. (2015). Role of halide anion on exciton binding energy and disorder in hybrid perovskite semiconductors. *2015 IEEE 42nd Photovolt. Spec. Conf. PVSC 2015*, 8–11, doi:10.1109/PVSC.2015.7355732.

38. D'Innocenzo, V., Grancini, G., Alcocer, M. J. P., Kandada, A. R. S., Stranks, S. D., Lee, M. M., Lanzani, G., Snaith, H. J., and Petrozza, A. (2014). Excitons versus free charges in organo-lead tri-halide perovskites. *Nat. Commun.*, **5**, 1–7.

39. Stranks, S. D., Burlakov, V. M., Leijtens, T., Ball, J. M., Goriely, A., and Snaith, H. J. (2014). Recombination kinetics in organic-inorganic perovskites: excitons, free charge, and subgap states. *Phys. Rev. Appl.*, **2**, 34007.

40. Hu, M., Bi, C., Yuan, Y., Xiao, Z., Dong, Q., Shao, Y., and Huang, J. (2015). Distinct exciton dissociation behavior of organolead trihalide perovskite and excitonic semiconductors studied in the same system. *Small*, **11**, 2164–2169.

41. Yamada, Y., Nakamura, T., Endo, M., Wakamiya, A., and Kanemitsu, Y. (2015). Photoelectronic responses in solution-processed perovskite $CH_3NH_3PbI_3$ solar cells studied by photoluminescence and photoabsorption spectroscopy. *IEEE J. Photovoltaics*, **5**, 401–405.

42. Cingolani, R., Calcagnile, L., Col?, G., Rinaldi, R., Lomoscolo, M., DiDio, M., Franciosi, a., Vanzetti, L., LaRocca, G. C., and Campi, D. (1996). Radiative recombination processes in wide-band-gap II–VI quantum wells: the interplay between excitons and free carriers. *J. Opt. Soc. Am. B*, **13**, 1268.

43. Saha, M. N. (1921). On a physical theory of stellar spectra. *Proc. R. Soc. Lond. A*, **99**, 135–153.

44. Miyata, K., Atallah, T. L., and Zhu, X. (2017). Lead halide perovskites : crystal-liquid duality, phonon glass electron crystals, and large polaron formation. *Sci. Adv.*, **3**, 1–10.

45. Yang, Y., Yang, M., Li, Z., Crisp, R., Zhu, K., and Beard, M. C. (2015). Comparison of recombination dynamics in $CH_3NH_3PbBr_3$ and $CH_3NH_3PbI_3$ perovskite films: influence of exciton binding energy. *J. Phys. Chem. Lett.*, **6**, 4688–4692.

46. Herz, L. M. (2016). Charge-carrier dynamics in organic-inorganic metal halide perovskites. *Annu. Rev. Phys. Chem.*, **67**, 65–89.

47. de Quilettes, D. W., Vorpahl, S. M., Stranks, S. D., Nagaoka, H., Eperon, G. E., Ziffer, M. E., Snaith, H. J., and Ginger, D. S. (2015). Impact of microstructure on local carrier lifetime in perovskite solar cells. *Science*, **348**, 683–686.

48. Xing, G., Wu, B., Wu, X., Li, M., Du, B., Wei, Q., Guo, J., Yeow, E. K. L., Sum, T. C., and Huang, W. (2017). Transcending the slow bimolecular recombination in lead-halide perovskites for electroluminescence. *Nat. Commun.*, **8**, 14558.

49. Stoumpos, C. C., Malliakas, C. D., and Kanatzidis, M. G. (2013). Semiconducting tin and lead iodide perovskites with organic cations: phase transitions, high mobilities, and near-infrared photoluminescent properties. *Inorg. Chem.*, **52**, 9019–9038.

50. Savenije, T. J., Ponseca, C. S., Kunneman, L., Abdellah, M., Zheng, K., Tian, Y., Zhu, Q., Canton, S. E., Scheblykin, I. G., Pullerits, T., Yartsev, A., and Sundström, V. (2014). Thermally activated exciton dissociation and recombination control the carrier dynamics in organometal halide perovskite. *J. Phys. Chem. Lett.*, **5**, 2189–2194.

51. Wehrenfennig, C., Liu, M., Snaith, H. J., Johnston, M. B., and Herz, L. M. (2014). Charge-carrier dynamics in vapour-deposited films of the organolead halide perovskite $CH_3NH_3PbI_{3-x}Cl_x$. *Energy Environ. Sci.*, **7**, 2269–2275.

52. Yamada, Y., Nakamura, T., Endo, M., Wakamiya, A., and Kanemitsu, Y. (2014). Photocarrier recombination dynamics in perovskite CH 3 NH 3 PbI 3 for solar cell applications. *J. Am. Chem. Soc.*, **136**, 11610–11613.

53. Noel, N. K., Abate, A., Stranks, S. D., Parrott, E. S., Burlakov, V. M., Goriely, A., and Snaith, H. J. (2014). Enhanced photoluminescence and solar cell performance via lewis base passivation of organic-inorganic lead halide perovskites. *ACS Nano*, **8**, 9815–9821.

54. Sear, R. P. (2007). Nucleation: theory and applications to protein solutions and colloidal suspensions. *J. Phys. Condens. Matter*, **19**, 33101.

55. Ahn, N., Kang, S. M., Lee, J.-W., Choi, M., and Park, N.-G. (2015). Thermodynamic regulation of $CH_3NH_3PbI_3$ crystal growth and its effect on photovoltaic performance of perovskite solar cells. *J. Mater. Chem. A*, **3**, 19901–19906.

56. Lewis, A. E., Zhang, Y., Gao, P., and Nazeeruddin, M. K. (2017). Unveiling the concentration-dependent grain growth of perovskite films from one- and two-step deposition methods: implications for photovoltaic application. *ACS Appl. Mater. Interfaces*, **9**, 25063–25066.

57. Bi, C., Wang, Q., Shao, Y., Yuan, Y., Xiao, Z., and Huang, J. (2015). Non-wetting surface-driven high-aspect-ratio crystalline grain growth for efficient hybrid perovskite solar cells. *Nat. Commun.*, **6**, 7747.

58. Kim, M. K., Jeon, T., Park, H. Il, Lee, J. M., Nam, S. A., and Kim, S. O. (2016). Effective control of crystal grain size in $CH_3NH_3PbI_3$ perovskite solar cells with a pseudohalide $Pb(SCN)_2$ additive. *CrystEngComm*, **18**, 6090–6095.

59. Ko, H., Sin, D. H., Kim, M., and Cho, K. (2017). Predicting the morphology of perovskite thin films produced by sequential deposition method: A crystal growth dynamics study. *Chem. Mater.*, **29**, 1165–1174.

60. Cho, H., Wolf, C., Kim, J. S., Yun, H. J., Bae, J. S., Kim, H., Heo, J.-M., Ahn, S., and Lee, T.-W. (2017). High-efficiency solution-processed inorganic metal halide perovskite light-emitting diodes. *Adv. Mater.*, **29**, 1700579.

61. Lee, S., Park, J. H., Lee, B. R., Jung, E. D., Yu, J. C., Di Nuzzo, D., Friend, R. H., and Song, M. H. (2017). Amine-based passivating materials for enhanced optical properties and performance of organic-inorganic perovskites in light-emitting diodes. *J. Phys. Chem. Lett.*, **8**, 1784–1792.

62. Byun, J., Cho, H., Wolf, C., Jang, M., Sadhanala, A., Friend, R. H., Yang, H., and Lee, T.-W. (2016). Efficient visible quasi-2D perovskite light-emitting diodes. *Adv. Mater.*, **28**, 7515–7520.

63. Yuan, M., Quan, L. N., Comin, R., Walters, G., Sabatini, R., Voznyy, O., Hoogland, S., Zhao, Y., Beauregard, E. M., Kanjanaboos, P., Lu, Z., Kim, D. H., and Sargent, E. H. (2016). Perovskite energy funnels for efficient light-emitting diodes. *Nat. Nanotechnol.*, **11**, 872–879.

64. Quan, L. N., Zhao, Y., García De Arquer, F. P., Sabatini, R., Walters, G., Voznyy, O., Comin, R., Li, Y., Fan, J. Z., Tan, H., Pan, J., Yuan, M., Bakr, O. M., Lu, Z., Kim, D. H., and Sargent, E. H. (2017). Tailoring the energy landscape in quasi-2D halide perovskites enables efficient green-light emission. *Nano Lett.*, **17**, 3701–3709.

65. Kerner, R. A., Zhao, L., Xiao, Z., and Rand, B. P. (2016). Ultrasmooth metal halide perovskite thin films via sol–gel processing. *J. Mater. Chem. A*, **4**, 8308–8315.

66. Zhao, X., Zhang, B., Zhao, R., Yao, B., Liu, X., Liu, J., and Xie, Z. (2016). Simple and efficient green-light-emitting diodes based on thin organolead bromide perovskite films via tuning precursor ratios and postannealing temperature. *J. Phys. Chem. Lett.*, **7**, 4259–4266.

67. Heo, J. H., Song, D. H., and Im, S. H. (2014). Planar $CH_3NH_3PbBr_3$ hybrid solar cells with 10.4% power conversion efficiency, fabricated by controlled crystallization in the spin-coating process. *Adv. Mater.*, **26**, 8179–8183.

68. Jeon, Y.-J., Lee, S., Kang, R., Kim, J.-E., Yeo, J.-S., Lee, S.-H., Kim, S.-S., Yun, J.-M., and Kim, D.-Y. (2015). Planar heterojunction perovskite solar cells with superior reproducibility. *Sci. Rep.*, **4**, 6953.
69. Yu, J. C., Kim, D. Bin, Jung, E. D., Lee, B. R., and Song, M. H. (2016). High-performance perovskite light-emitting diodes via morphological control of perovskite films. *Nanoscale*, **8**, 7036–7042.
70. Eperon, G. E., Stranks, S. D., Menelaou, C., Johnston, M. B., Herz, L. M., and Snaith, H. J. (2014). Formamidinium lead trihalide: a broadly tunable perovskite for efficient planar heterojunction solar cells. *Energy Environ. Sci.*, **7**, 982.
71. Jeon, Y.-J., Lee, S., Kang, R., Kim, J.-E., Yeo, J.-S., Lee, S.-H., Kim, S.-S., Yun, J.-M., and Kim, D.-Y. (2015). Planar heterojunction perovskite solar cells with superior reproducibility. *Sci. Rep.*, **4**, 6953.
72. Chausov, F. F. (2008). Effect of adsorbed impurities on the crystal growth of poorly soluble salts from slightly supersaturated solutions. *Theor. Found. Chem. Eng.*, **42**, 179–186.
73. Di, D., Musselman, K. P., Li, G., Sadhanala, A., Ievskaya, Y., Song, Q., Tan, Z. K., Lai, M. L., MacManus-Driscoll, J. L., Greenham, N. C., and Friend, R. H. (2015). Size-dependent photon emission from organometal halide perovskite nanocrystals embedded in an organic matrix. *J. Phys. Chem. Lett.*, **6**, 446–450.
74. Meng, F., Zhang, C., Chen, D., Zhu, W., Yip, H.-L., and Su, S.-J. (2017). Combined optimization of emission layer morphology and hole-transport layer for enhanced performance of perovskite light-emitting diodes. *J. Mater. Chem. C*, **5**, 6169–6175.
75. Li, G., Tan, Z., Di, D., Lai, M. L., Jiang, L., Lim, J. H., Friend, R. H., and Greenham, N. C. (2015). Efficient light-emitting diodes based on nanocrystalline perovskite in a dielectric polymer matrix. *Nano Lett.*, **15**, 2640–2644.
76. Wu, C., Zou, Y., Wu, T., Ban, M., Pecunia, V., Han, Y., Liu, Q., Song, T., Duhm, S., and Sun, B. (2017). Improved performance and stability of all-inorganic perovskite light-emitting diodes by antisolvent vapor treatment. *Adv. Funct. Mater.*, **27**, 1700338.
77. Li, J., Shan, X., Bade, S. G. R., Geske, T., Jiang, Q., Yang, X., and Yu, Z. (2016). Single-layer halide perovskite light-emitting diodes with sub-band gap turn-on voltage and high brightness. *J. Phys. Chem. Lett.*, **7**, 4059–4066.
78. Li, J., Cai, F., Yang, L., Ye, F., Zhang, J., Gurney, R. S., Liu, D., and Wang, T. (2017). Sodium bromide additive improved film morphology and

performance in perovskite light-emitting diodes. *Appl. Phys. Lett.*, **111**, 53301.

79. La-Placa, M.-G., Longo, G., Babaei, A., Martínez-Sarti, L., Sessolo, M., and Bolink, H. J. (2017). Photoluminescence quantum yield exceeding 80% in low dimensional perovskite thin-films via passivation control. *Chem. Commun.*, **53**, 8707–8710.
80. Hu, H., Salim, T., Chen, B., and Lam, Y. M. (2016). Molecularly engineered organic-inorganic hybrid perovskite with multiple quantum well structure for multicolored light-emitting diodes. *Sci. Rep.*, **6**, 33546.
81. Zhang, S., Yi, C., Wang, N., Sun, Y., Zou, W., Wei, Y., Cao, Y., Miao, Y., Li, R., Yin, Y., Zhao, N., Wang, J., and Huang, W. (2017). Efficient red perovskite light-emitting diodes based on solution-processed multiple quantum wells. *Adv. Mater.*, **29**, 1606600.
82. Wei, M., Sun, W., Liu, Y., Liu, Z., Xiao, L., Bian, Z., and Chen, Z. (2016). Highly luminescent and stable layered perovskite as the emitter for light emitting diodes. *Phys. Status Solidi Appl. Mater. Sci.*, **213**, 2727–2732.
83. Wei, Z., Perumal, A., Su, R., Sushant, S., Xing, J., Zhang, Q., Tan, S. T., Demir, H. V., and Xiong, Q. (2016). Solution-processed highly bright and durable cesium lead halide perovskite light-emitting diodes. *Nanoscale*, **8**, 18021–18026.
84. Kumawat, N. K., Dey, A., Narasimhan, K. L., and Kabra, D. (2015). Near infrared to visible electroluminescent diodes based on organometallic halide perovskites: structural and optical investigation. *ACS Photonics*, **2**, 349–354.
85. Edri, E., Kirmayer, S., Kulbak, M., Hodes, G., and Cahen, D. (2014). Chloride inclusion and hole transport material doping to improve methyl ammonium lead bromide perovskite-based high open-circuit voltage solar cells. *J. Phys. Chem. Lett.*, **5**, 429–433.
86. Kumar, P., Zhao, B., Friend, R. H., Sadhanala, A., and Narayan, K. S. (2017). Kinetic control of perovskite thin-film morphology and application in printable light-emitting diodes. *ACS Energy Lett.*, **2**, 81–87.
87. Xiao, M., Huang, F., Huang, W., Dkhissi, Y., Zhu, Y., Etheridge, J., Gray-Weale, A., Bach, U., Cheng, Y.-B., and Spiccia, L. (2014). A fast deposition-crystallization procedure for highly efficient lead iodide perovskite thin-film solar cells. *Angew. Chem.*, **126**, 10056–10061.
88. Jeon, N. J., Noh, J. H., Kim, Y. C., Yang, W. S., Ryu, S., and Seok, S. Il. (2014). Solvent engineering for high-performance inorganic–organic hybrid perovskite solar cells. *Nat. Mater.*, **13**, 897–903.

89. Rong, Y., Tang, Z., Zhao, Y., Zhong, X., Venkatesan, S., Graham, H., Patton, M., Jing, Y., Guloy, A. M., and Yao, Y. (2015). Solvent engineering towards controlled grain growth in perovskite planar heterojunction solar cells. *Nanoscale*, **7**, 10595–10599.

90. Zhou, Y., Game, O. S., Pang, S., and Padture, N. P. (2015). Microstructures of organometal trihalide perovskites for solar cells: their evolution from solutions and characterization. *J. Phys. Chem. Lett.*, **6**, 4827–4839.

91. Salim, T., Sun, S., Abe, Y., Krishna, A., Grimsdale, A. C., and Lam, Y. M. (2015). Perovskite-based solar cells: impact of morphology and device architecture on device performance. *J. Mater. Chem. A*, **3**, 8943–8969.

92. Kim, M. K., Jeon, T., Park, H. Il, Lee, J. M., Nam, S. A., and Kim, S. O. (2016). Effective control of crystal grain size in $CH_3NH_3PbI_3$ perovskite solar cells with a pseudohalide $Pb(SCN)_2$ additive. *CrystEngComm*, **18**, 6090–6095.

93. De Bastiani, M., D'Innocenzo, V., Stranks, S. D., Snaith, H. J., and Petrozza, A. (2014). Role of the crystallization substrate on the photoluminescence properties of organolead mixed halides perovskites. *APL Mater.*, **2**.

94. Kim, D. Bin, Yu, J. C., Nam, Y. S., Kim, D. W., Jung, E. D., Lee, S. Y., Lee, S., Park, J. H., Lee, A.-Y., Lee, B. R., Di Nuzzo, D., Friend, R. H., and Song, M. H. (2016). Improved performance of perovskite light-emitting diodes using a PEDOT:PSS and MoO_3 composite layer. *J. Mater. Chem. C*, **4**, 8161–8165.

95. Wang, J., Wang, N., Jin, Y., Si, J., Tan, Z. K., Du, H., Cheng, L., Dai, X., Bai, S., He, H., Ye, Z., Lai, M. L., Friend, R. H., and Huang, W. (2015). Interfacial control toward efficient and low-voltage perovskite light-emitting diodes. *Adv. Mater.*, **27**, 2311–2316.

96. Zuo, L., Gu, Z., Ye, T., Fu, W., Wu, G., Li, H., and Chen, H. (2015). Enhanced photovoltaic performance of $CH_3NH_3PbI_3$ perovskite solar cells through interfacial engineering using self-assembling monolayer. *J. Am. Chem. Soc.*, **137**, 2674–2679.

97. Chih, Y. K., Wang, J. C., Yang, R. T., Liu, C. C., Chang, Y. C., Fu, Y. S., Lai, W. C., Chen, P., Wen, T. C., Huang, Y. C., Tsao, C. S., and Guo, T. F. (2016). NiOx electrode interlayer and $CH_3NH_2/CH_3NH_3PbBr_3$ interface treatment to markedly advance hybrid perovskite-based light-emitting diodes. *Adv. Mater.*, **28**, 8687–8694.

98. Meng, L., Yao, E.-P., Hong, Z., Chen, H., Sun, P., Yang, Z., Li, G., and Yang, Y. (2017). Pure formamidinium-based perovskite light-emitting diodes with high efficiency and low driving voltage. *Adv. Mater.*, **29**, 1603826.

99. Kumawat, N. K., Jain, N., Dey, A., Narasimhan, K. L., and Kabra, D. (2017). Quantitative correlation of perovskite film morphology to light emitting diodes efficiency parameters. *Adv. Funct. Mater.*, **27**, 1603219.

100. Cheng, Y., Li, H.-W., Zhang, J., Yang, Q.-D., Liu, T., Guan, Z., Qing, J., Lee, C.-S., and Tsang, S.-W. (2016). Spectroscopic study on the impact of methylammonium iodide loading time on the electronic properties in perovskite thin films. *J. Mater. Chem. A*, **4**, 561–567.

Index

CPSIA information can be obtained
at www.ICGtesting.com
Printed in the USA
BVHW010003170422
634394BV00002B/28